T0253922

Semantics in Mobile Sensing

Synthesis Lectures on the Semantic Web: Theory and Technology

Editor
James Hendler, *Rensselaer Polytechnic Institute*
Ying Ding, *Indiana University*

Whether you call it the Semantic Web, Linked Data, or Web 3.0, a new generation of Web technologies is offering major advances in the evolution of the World Wide Web. As the first generation of this technology transitions out of the laboratory, new research is exploring how the growing Web of Data will change our world. While topics such as ontology-building and logics remain vital, new areas such as the use of semantics in Web search, the linking and use of open data on the Web, and future applications that will be supported by these technologies are becoming important research areas in their own right. Whether they be scientists, engineers or practitioners, Web users increasingly need to understand not just the new technologies of the Semantic Web, but to understand the principles by which those technologies work, and the best practices for assembling systems that integrate the different languages, resources, and functionalities that will be important in keeping the Web the rapidly expanding, and constantly changing, information space that has changed our lives.

Topics to be included:

- Semantic Web Principles from linked-data to ontology design

- Key Semantic Web technologies and algorithms

- Semantic Search and language technologies

- The Emerging "Web of Data" and its use in industry, government and university applications

- Trust, Social networking and collaboration technologies for the Semantic Web

- The economics of Semantic Web application adoption and use

- Publishing and Science on the Semantic Web

- Semantic Web in health care and life sciences

Semantics in Mobile Sensing
Zhixian Yan and Dipanjan Chakraborty
2014

Provenance: An Introduction to PROV
Luc Moreau and Paul Groth
2013

Resource-Oriented Architecture Patterns for Webs of Data
Brian Sletten
2013

Aaron Swartz's A Programmable Web: An Unfinished Work
Aaron Swartz
2013

Incentive-Centric Semantic Web Application Engineering
Elena Simperl, Roberta Cuel, and Martin Stein
2013

Publishing and Using Cultural Heritage Linked Data on the Semantic Web
Eero Hyvönen
2012

VIVO: A Semantic Approach to Scholarly Networking and Discovery
Katy Börner, Michael Conlon, Jon Corson-Rikert, and Ying Ding
2012

Linked Data: Evolving the Web into a Global Data Space
Tom Heath and Christian Bizer
2011

© Springer Nature Switzerland AG 2022
Reprint of original edition © Morgan & Claypool 2014

All rights reserved. No part of this publication may be reproduced, stored in a retrieval system, or transmitted in any form or by any means—electronic, mechanical, photocopy, recording, or any other except for brief quotations in printed reviews, without the prior permission of the publisher.

Semantics in Mobile Sensing
Zhixian Yan and Dipanjan Chakraborty

ISBN: 978-3-031-79452-0   paperback
ISBN: 978-3-031-79453-7   ebook

DOI 10.1007/978-3-031-79453-7

A Publication in the Springer series
SYNTHESIS LECTURES ON THE SEMANTIC WEB: THEORY AND TECHNOLOGY

Lecture #8
Series Editors: James Hendler, *Rensselaer Polytechnic Institute*
                Ying Ding, *Indiana University*
Series ISSN
Print 2160-4711   Electronic 2160-472X

Semantics in Mobile Sensing

Zhixian Yan
École Polytechnique Fédérale de Lausanne (EPFL)

Dipanjan Chakraborty
IBM Research India

*SYNTHESIS LECTURES ON THE SEMANTIC WEB: THEORY AND
TECHNOLOGY #8*

ABSTRACT

The dramatic progress of smartphone technologies has ushered in a new era of mobile sensing, where traditional wearable on-body sensors are being rapidly superseded by various embedded sensors in our smartphones. For example, a typical smartphone today, has at the very least a GPS, WiFi, Bluetooth, triaxial accelerometer, and gyroscope. Alongside, new accessories are emerging such as proximity, magnetometer, barometer, temperature, and pressure sensors. Even the default microphone can act as an acoustic sensor to track noise exposure for example. These sensors act as a "lens" to understand the user's context along different dimensions.

Data can be passively collected from these sensors without interrupting the user. As a result, this new era of mobile sensing has fueled significant interest in understanding what can be extracted from such sensor data both instantaneously as well as considering volumes of time series from these sensors. For example, GPS logs can be used to determine automatically the significant places associated to a user's life (e.g., home, office, shopping areas). The logs may also reveal travel patterns, and how a user moves from one place to another (e.g., driving or using public transport). These may be used to proactively inform the user about delays, relevant promotions from shops, in his "regular" route. Similarly, accelerometer logs can be used to measure a user's average walking speed, compute step counts, gait identification, and estimate calories burnt per day. The key objective is to provide better services to end users.

The objective of this book is to inform the reader of the methodologies and techniques for extracting meaningful information (called "semantics") from sensors on our smartphones. These techniques form the cornerstone of several application areas utilizing smartphone sensor data. We discuss technical challenges and algorithmic solutions for modeling and mining knowledge from smartphone-resident sensor data streams. This book devotes two chapters to dive deep into a set of highly available, commoditized sensors—the positioning sensor (GPS) and motion sensor (accelerometer). Furthermore, this book has a chapter devoted to energy-efficient computation of semantics, as battery life is a major concern on user experience.

KEYWORDS

semantic analytics, smartphone, mobile sensing, crowdsourcing, people-centric sensing, semantic trajectories, trajectory ontologies, semantic activities, activity recognition, energy-efficient computation

Contents

Acknowledgments xi

1 Introduction 1
  1.1 Mobile Sensing: Definitions and Scope 1
    1.1.1 Smartphone-based Sensing 2
    1.1.2 Sensing Architectures 5
  1.2 Semantics from Sensors 7
    1.2.1 Semantic Modeling of Mobile Sensors 8
    1.2.2 Semantic Computation from Mobile Sensors 9
  1.3 Book Structure 10

2 Semantic Trajectories from Positioning Sensors 11
  2.1 The Positioning Sensor 12
    2.1.1 Sensor Type and Functionality 12
    2.1.2 Dealing with Positioning Data: Technical Summary 14
    2.1.3 From GPS to Semantic Trajectories 16
  2.2 Semantic Trajectory Modeling 16
    2.2.1 Semantic Trajectory Ontology 17
    2.2.2 Hybrid Semantic Trajectory Model 22
  2.3 Semantic Trajectory Computation 26
    2.3.1 Data Preprocessing Layer 27
    2.3.2 Trajectory Identification Layer 31
    2.3.3 Trajectory Structure Layer 33
  2.4 Semantic Trajectory Annotation 37
    2.4.1 Annotation with Semantic Regions 40
    2.4.2 Annotation with Semantic Lines 41
    2.4.3 Annotation with Semantic Points 44
  2.5 Summary and Outlook 48

3 Semantic Activities from Motion Sensors 51
  3.1 The Motion Sensor 52
    3.1.1 Sensor Functionality 52

		3.1.2	What can we Learn From Motion?	53
		3.1.3	Data Collection	55
	3.2	Feature Spaces		57
		3.2.1	Data Processing	57
		3.2.2	Feature Classes—An Overview	58
		3.2.3	Feature Computation and Energy	61
	3.3	Activity Learning Techniques		62
		3.3.1	Learning Models	63
		3.3.2	Choice of Learning Models	67
		3.3.3	Testing	67
	3.4	Case Study: Micro Activity (MA) Learning		68
		3.4.1	Data Collection	68
		3.4.2	Feature Vector Representation	69
		3.4.3	Results of MA Supervised Learning	70
	3.5	Case Study: Complex Activity Learning		71
		3.5.1	User Recruitment and Data Collection	72
		3.5.2	Data Processing & Sanitization	73
		3.5.3	Features for Complex Activities	73
		3.5.4	Complex Activity Learning Approaches	78
		3.5.5	Learning Performance	80
	3.6	Conclusions and Summary		82
4	**Energy-Efficient Computation of Semantics from Sensors**			**85**
	4.1	Energy-Efficient Mobile Sensing		85
		4.1.1	Smartphone Battery Limitations	85
		4.1.2	Energy-Efficient Sensing: Hardware Approaches	87
		4.1.3	Energy-Efficient Sensing: Software Approaches	87
	4.2	Model-based Energy-Efficient Sensing		89
		4.2.1	Two-Tier Optimal Sensing	91
		4.2.2	Model-Based Optimal Segmentation	93
		4.2.3	Model-Based Optimal Sensor Sampling	97
	4.3	Energy-Efficient Semantic Activity Learning		99
		4.3.1	Classification Accuracy vs. Energy Consumption	100
		4.3.2	A3R—A methodology for continuous adaptive sampling	105
	4.4	Concluding Remarks		107

5   Conclusion **109**

    5.1   Summary 109

    5.2   Challenges and Opportunities 111

Bibliography **115**

Authors' Biographies **131**

Acknowledgments

We would like to express our deepest gratitude to a few excellent researchers and supporters, without whom the book would not exist. Much of the technical work behind this book was conducted when both authors were affiliated with EPFL—Zhixian was first a Ph.D. and then a PostDoc researcher at EPFL, Dipanjan was a visiting researcher. The mobile sensing work was supervised by Prof. Karl Aberer who runs the Distributed Information Systems group at EPFL. We are grateful to Prof. Aberer for his guidance and gracious support to initiate projects in mobile sensing, which formed the fundamental building blocks of this book. This work was sponsored by several Swiss or European research projects at EPFL, including NCCR MICS, OpenSense, OpenIOT, and MODAP.

We owe special thanks to Prof. Stefano Spaccapietra and Prof. Christine Parent at EPFL. They contributed significantly to the key concepts of the semantic trajectories in Chapter 2. Prof. Spaccapietra and Prof. Parent are the first researchers to have initiated the semantic trajectory area, and offered us tremendous support.

Several other researchers and colleagues assisted us in writing this book. Prof. Archan Misra from Singapore Management University researched with us on the topics of activity recognition, which has substantial contributions in Chapter 3. Julien Eberle at EPFL worked with us on collaborative sensing, which becomes a key part of Chapter 4 on energy-efficient mobile sensing. We are thankful to Parikshit Sharma, Sheetal Agarwal, Saket Sathe, and Deeptish Mukherjee who assisted us with reviews and comments.

We would like to acknowledge some of our close family members who have supported us and encouraged us throughout. Dipanjan would like to thank his wife Laura and his parents and in-laws, Joydeb, Nupur, Alexis, and Leela, for their unending love, encouragement, and guidance. Zhixian, in particular, wants to thank his wife Jing and his newborn son Bokai for their patience, support and making the writing of this book a pleasant experience.

We are grateful to the three editors, i.e., Michael B. Morgan (President and CEO of Morgan & Claypool Publishers), Prof. Ying Ding from Indiana University, and Prof. Jim Hendler from Rensselaer Polytechnic Institute for keeping us on track as well as for their active involvement in making this book happen.

Zhixian Yan and Dipanjan Chakraborty
April 2014

CHAPTER 1

Introduction

1.1 MOBILE SENSING: DEFINITIONS AND SCOPE

The term *Mobile Sensing* is not new. Since the late 1980's and early 1990's, one branch of mobile and wireless computing has focused on creating various sensor equipment and establishing sensor networks to monitor phenomena of interest [3, 4] in many applications, such as monitoring atmosphere [45], odor measurement via gas sensor [81], and potholes on road surface [47]. Since then, after the leapfrog of wireless networking technologies, sensors started becoming increasingly connected to each other, or to backend servers via different networking technologies. Usually, each sensor motherboard has a *sensing* module and a *communications* module.

As sensors started becoming untethered, researchers focused on how to increase coverage and reliability of sensing while sensors are moving around. For a long while, traditional wireless network research has focused on different inter-related issues associated with mobile sensors. For example, one rich area of research investigated how to place sensors in an area in order to maximize coverage [25, 39, 78]. Another area looked at how to generate different types of mobility models that can be used by the sensors as "mobility policies" to guide how the sensors should be moving around in an area. A very popular, although old, example is that of random waypoint mobility pattern [22]. The primary application area for this body of work is in battle field surveillance missions, space and land surveillance, deep sea monitoring, etc. Here, the sensor network (i.e., the network of sensor nodes) talk to each other to determine how to best operate, what type of data to collect, when to transmit data to the backend, etc. An overarching need in this body of work is to reduce the energy consumed by the sensors in performing their operations. Energy is a scarce resource, especially in situations where the node is dependent on a battery for its survival.

While the wireless sensor network community focused on the above application areas and associated technical problems, another body of research work started surfacing up around the mid 1990's. This was called *wearable computing*. In this book, the primary objective was to monitor and map movement patterns of living organisms—animals, birds, humans, etc.—using a plethora of sensors that are worn or carried with the organism. More often than not, the primary direction of research in wearable computing has been *inward*, i.e., to use sensors to make observations about the carrier, rather than to sense a phenomenon of interest (e.g., environment). The observations made have been used in innovative ways to make inferences about the carrier [108, 128] and to establish argument reality [149]. For example, on-body acceleration sensors, worn in different parts, can be used to figure out locomotory movements, location sensors can be used to figure out mobility patterns, and RFID badges can be used to figure out proximity to certain surroundings, e.g., Bluetooth enabled "smart objects" in a smart room.

The line between *mobile* (*moving object*) *sensor networks* and *wearable computing* started getting blurred around the mid 2000's with the advent of smartphones. Smartphones come with a number of special-purpose sensors (e.g., GPS, Accelerometer, Gyroscope) or communication units (e.g., Bluetooth, WiFi). The output trace from these sensors or communication units reveal rich information about the smartphone, as a new paradigm of mobile phone sensing [96], which is also called *people-centric sensing* [23]. A smartphone can be easily seen as a powerful computing and communication device that can be used with traditional sensors (e.g., temperature sensor) to report statistics about underlying environment. For the traditional sensor networking community, a smartphone presents a powerful, pervasive, and well-adopted computing and communication tool that can be used in many different ways. For the wearable computing community, the smartphone represents a powerful wearable multi-sensor unit that can be used to infer knowledge about the carrier as well as the underlying surrounding. With the rapid growth of cellular data and network standards, we are witnessing an improvement in the data bandwidth available with us. Scientists are of the belief that this combined trend of increased network bandwidth and speed, coupled with the advanced sensing and processing capabilities of smartphones and its ability to operate as a *sensor docking station*, is going to change the landscape of mobile sensing in the future.

This book goes beyond the contemporary meaning of the term mobile sensing and investigates semantic data extraction in an era where traditional sensor networks have united with wearable sensing. In particular, we are largely going to focus on the area encompassing (1) embedded or extended sensors that are on the smartphone, (2) carriers that are usually people or community-driven objects such as cars, buses, etc., and (3) mobility that is usually un-organized and unstructured and bottom-up, i.e., the community autonomously governs the mobility. Hence, our scope of the term *mobile sensing* is going to largely represent the above real-life conditions.

In the next subsection, we provide a high-level background of smartphone sensing, sensing architectures, and trends.

1.1.1   SMARTPHONE-BASED SENSING

Andrew Campbell from Darthmouth College, Hanover, New Hampshire, USA, is one of the first researchers to have coined the term "people sensing" back in 2005 [24]. The word meant that traditional sensor networks are going to slowly evolve into a model where people will be carrying sensing devices and using personal devices to sense heterogeneous phenomenon of interest—in particular, on the applications of urban sensing. However, in 2005, smartphones were still difficult to program. The capabilities in terms of processing power and on-board sensing units were also limited. Since then, over the last eight years, significant progress has been made in the market. The introduction of the iPhone OS[1] in 2007 and the Android[2] platform in 2008 opened up the

---

[1]http://www.apple.com/iphone/ios/
[2]http://www.android.com/

operating system for rapid and easy programming and community-driven development. Progressively with time, more and more cheap and easy-to-program sensors started getting embedded into the phone. The phone that we carry, slowly but steadily, started moving from a perception of *communication device* to a *personal intelligent assistant*.

The smartphone today has several on-board sensing units. Many physical sensors are those units that explicitly measure certain attribute of the environment (e.g., location, temperature). Other sensors can be those whose primary functionality is not to sense, but the data implicitly contains some traces of the environment. An example of an "explicit" physical sensor is the Global Positioning System (GPS) sensor. An example of an "implicit" sensor is Bluetooth. Bluetooth is primarily used for peer-to-peer communications, but the device discovery module of Bluetooth can periodically report the other Bluetooth-enabled devices that can be observed in its vicinity. This can be used to infer some context about the surroundings of the device. We provide a typical list of physical sensors that come embedded with some of the latest smartphone handsets today.

- **Global Positioning System—GPS.** This is one of the most well-known sensors on the smartphone. It computes the position of the smartphone using satellites and their ground stations as a reference frame. The output is typically a ⟨*latitude, longitude, altitude*⟩ tuple, along with other information such as velocity and direction. An extensive tutorial on GPS is provided by Trimble,[3] and more detailed explanations of GPS can be found in [125].

- **Accelerometer—ACC.** An accelerometer records acceleration along three mutually perpendicular axes x, y, z. When the smartphone is placed horizontal on a table, one axis points towards gravity ($\vec{g}$) vector. Accelerometers play an active role in measuring physical activities [156]. They are typically used to detect the orientation of a smartphone and widely used today in various gaming and activity recognition applications [139, 170]. Readers interested in knowing physical properties of a typical accelerometer chip may refer to wikipedia.[4]

- **Gyroscope—GYR.** A gyroscope is used to measure orientation of the smartphone using principles of angular momentum. It can measure the rate of change of angle along a particular axis and is typically reliable to detect short-term and bursty orientation changes accurately. Smartphones typically use a combination of accelerometer and gyroscope to calibrate each other [37] as well as reliably detect orientation changes, locomotive states, etc. [107]

- **Bluetooth—BT.** Bluetooth is used to exchange data over short distances between two peers using radio transmissions in the ISM band of 2400–2480 Mhz. Bluetooth can hence discover other peer Bluetooth-enabled devices around itself. Thanks to the new Bluetooth LE (Low Energy) technology, smartphones can more efficiently capture proximity data. This property has been widely used in research to estimate and sense real-time social surroundings, e.g., estimating the population density [161], computing the groups [41], and inferring social contexts [142].

[3]http://www.trimble.com/gps_tutorial/howgps.aspx
[4]http://en.wikipedia.org/wiki/Accelerometer

- **Light—LT.** The light sensors are typically photodiodes measuring light intensity, by correlating it with measured current. Light sensors are widely used to control screen brightness in different luminous settings. It is also used to lock the touchscreen when a user is holding the phone against his ear. Light sensor is also integrated with other sensors for intelligent services such as better localization with the combination of sound [7] and phone position detection (e.g., in pocket, on table, in hand) together with accelerometer [144].

The list of phone sensors does not end here. There are other sensors such as NFC-related proximity sensors and barometers. Of course, the camera and the microphone—two most prevalent components can also act as sources to feed information about a smartphone's surrounding, visual, and acoustic context. Apart from these embedded sensors, smartphones come with input/output interfaces (like micro-USB ports) that can be used by many accessory manufacturers to plug-in accessories like a headset, etc. Recently, a startup built accessory that plugs into the mobile devices' (iPhone or Android devices) audio jack for flexible credit card payment using Square.[5] Researchers and practitioners are investing money to prepare novel accessories which can be used as add-on gadgets along with a smartphone.

Each of the above sensors have specific utility for their inclusion in the handset or as an accessory. For example, GPS is used to supply real-time location of the device to map applications (such as Google maps), which can help the user to localize herself in unknown surroundings, find directions, etc. An accelerometer measures the acceleration of the device, typically along three dimensions—along the vector pointing towards the center of earth (called the gravity vector), and two vectors perpendicular to the gravity vector. Accelerometer is often used as a motion sensor, for stabilization of the camera during videography and photography. It is also used to measure the orientation of the phone. Similarly, proximity and light sensor is often used to control the illumination intensity of the screen and save on power.

While the first wave of applications of these sensors is primarily to drive user experience (e.g., to change display orientation by analyzing real-time accelerometer data), researchers across the globe are focusing on a subsequent wave of innovative applications, exploiting these smartphone sensors to capture, analyze, and predict several environmental properties of the users. Lane et al., in their survey on *Mobile Phone Sensing* in 2010 [96], observed "Now phones can be programmed to support new disruptive sensing applications such as sharing the user's real-time activity with friends on social networks such as Facebook, keeping track of a person's carbon footprint, or monitoring a user's well being. Second, smartphones are open and programmable. In addition to sensing, phones come with computing and communication resources that offer a low barrier of entry for third-party programmers (e.g., undergraduates with little phone programming experience are developing and shipping applications). Third, importantly, each phone vendor now offers an app store allowing developers to deliver new applications to large populations of users across the globe, which is transforming the deployment of new applications, and allowing the collection and analysis of data far beyond the scale of what was previously possible. Fourth, the mobile

---

[5]https://squareup.com/

computing cloud enables developers to offload mobile services to back-end servers, providing unprecedented scale and additional resources for computing on collections of large-scale sensor data and supporting advanced features such as persuasive user feedback based on the analysis of big sensor data."

In [154], the authors categorize people-driven sensing using physical sensors in terms of three types of observation properties: (1) spatio-temporal properties (e.g., presence, count, location, track, identity); (2) behavioral properties (e.g., pose, action, activity, behavior, group behavior); and (3) physiological properties (e.g., temperature, blood pressure, heart rate). In addition to sensing these personalized properties, sensors also capture information of the nature, e.g., to monitor environment like air quality and climate change, as well as their consequences [29]. Many novel sensing applications have been emerging, such as using mobile phones to detect earthquakes [51], deploying sensors on public transports for air quality monitoring [2], and the Copenhagen Wheel for promptly analyzing pollution levels, traffic congestion, and road conditions using e-bikes.[6]

It is worth noting that there is more emerging interest in studying soft-sensors rather than previously mentioned physical sensors. For example, human can easily report massive real-time social data using the rapidly growing social networks, e.g., Facebook, Twitter, YouTube, and location-based social networking services like Foursquare check-in. Therefore, in the "human as sensors" paradigm, a sensor is not necessarily a hardware sensor but also a virtual sensor also-called a "social sensor" or "logical/soft sensor" in literature. Such virtual sensing in social networks allows efficient and effective information sharing and propagation, with an unprecedented deployment scale, which in turn open novel data-driven applications in numerous domains such as health, transportation, energy, disaster recovery, and even warfare [147]. These applications are often referred to as "crowd-sourcing" or "citizen sensing" scenarios, since they are capitalizing on the power of crowds and relying on a large scale of citizens participation [17]. In addition to using public social networks/media like Twitter as virtual sensors in building crowdsourcing applications, several dedicated open-source platforms are alternatively built for easy crowdsourcing development, e.g., GeoCha[7] and Ushahidi.[8] It is worth noting that user reports via virtual sensors would be biased and subjective, but we can use that fact to identify communities or group and do relevant data filtering. In this book, the main focus is on studying the semantic computation over physical sensors, not soft sensors.

1.1.2   SENSING ARCHITECTURES

Information sensed by the users and their smartphones may be transmitted to a back-end server. Information from multiple devices and users may be combined together to reveal significant trends of an environment like predicting air quality, crowd levels, and generally, information

---

[6]http://senseable.mit.edu/copenhagenwheel/
[7]http://instedd.org/technologies/geochat/
[8]http://ushahidi.com/

relating several important city management sectors like traffic issues, neighborhood issues, emergency, civic complaints, etc. These developments are known under the names of "urban sensing" [36, 95], "participatory sensing" [21, 63], "opportunistic sensing" [85], "community sensing" [90], "crowdsensing" [60], "crowdsourcing" [79, 89], "people sensing" [24], etc. These buzz words all describe the space of sensing architectures from various application perspectives. Nevertheless, the key point from all buzz words is to build applications that can benefit from participators or volunteers of a large number of people with mobile devices. The data/contributions from participators can be generated not only actively but also passively. The fundamental difference between this new paradigm of community-driven sensing and traditional sensor networks is in its organic involvement model, and autonomous, human driven nature of the sensor network.

Among these buzz words, *crowdsensing* and *crowdsourcing* are largely used, in recent years, for describing such community-driven mobile sensing designs. Crowdsensing is a relatively new term compared to Crowdsourcing. The fundamental difference between crowdsourcing and crowdsensing is in the nature of the task. Typically, crowdsourcing tasks are top-down driven, i.e., a task is decomposed into suitable pieces that can be executed by human endpoints. Thereafter, different solutions are used to reach out to the community and get the pieces executed. Crowdsensing, on the other hand, is a bottom-up driven method, i.e., the community is organically involved in sensing a phenomenon of interest. The basic blueprint of a crowdsensing application consists of a sensing agent running on the phone, while a back-end server aggregates results and provides services.

In terms of on-the-field deployments, there are quite a few trends. With the growing number of cities and populations within them, continuous monitoring of city's infrastructure for growth and sustainable development is gaining more and more importance. Increasingly, the city is being instrumented and inter-wired with millions of heterogeneous sensing infrastructures (traffic surveillance, CCTV footages in popular areas, sensors in malls, indoor localization). In parallel, the community is increasingly starting to play a role in monitoring urban dynamics using their smartphones, powered by sensors and social media tools. Therefore, the scope of "human as sensors" extend beyond the embedded sensors and also include what can be sensed by a user's biological sense organs. Social media offers a "channel" for capturing such sensed data about a city, its traffic, social and public events, etc. Social media data collection infrastructures, combined with a handy smartphone with several social update applications, provide an easy-to-use mechanism for users to provide just-in-time updates, and capture unprecedented data about a city. For example, we already know how social media assisted in the Spring uprisings in Africa. Each of these social media infrastructures largely follow a client–server oriented architecture, where a thin client is used to provide updates to a user and receive inputs. There is a significant amount of work happening in the social media space to glean relevant knowledge from this high volume, noisy, text data stream [74, 84].

The concept of "humans as sensors" to observe, infer, and predict properties of our environment has gained significant attention over the last few years. For example, GPS trajectories generated by running the GPS sensor periodically from smartphones can be used to understand urban densities and population dynamics. At a personal level, such trajectories contain rich information of our day-to-day life and can be used in several ways to provide us with meaningful recommendations, exploit activities and activity logs, and better understand health quality. Similarly, accelerometer can be used to monitor the amount of time we are walking or performing fitness routines. Research has made significant progress on understanding algorithms—methodologies for knowledge extraction from these sensor streams. However, quality and reliability of predictions "in the wild," i.e., as the user moves around in day-to-day life is an open question. It has been studied that laboratory-constrained environments and results obtained via such experiments are not synergistic with experiments in the wild. In this book, we will provide the reader with a flavor of algorithms and techniques that are used for knowledge extraction from such smartphone sensor data streams, and provide the reader with inputs and insights on open issues and challenges of applying these in real-world settings. The learning and inference should be robust and domain specific.

In this book, we would discuss and present the emergent architectures that are being proposed to capture data and extract knowledge or semantics out of such city-scale deployments. In particular, we focus on the main techniques for inferring meaningful information from sensor data captured through mobile phones towards such a model of crowdsensing. In addition, we present energy-efficient sensing approaches that are highly applicable in smartphone platforms.

1.2   SEMANTICS FROM SENSORS

The word *semantics* popularly means "the study of meanings." In linguistics, it is used to represent the interpretation of words, phrases, signs and symbols in a given context (i.e., a bag of such words, phrases, symbols) [150]. Compared with its extensive study in the area of Web and natural language processing, in particular with the booming of Semantic Web technologies in recent years, *semantics* is much less studied in the mobile sensing area. For mobile sensing, it's non-trivial to provide a universal definition for semantics; but, in general, *semantics* in mobile sensing very broadly refers to any meaningful knowledge that we can extract from the sensor data. For example, GPS sensor provides a geometric location point with the ⟨*latitude, longitude*⟩ coordinates. The semantics of this point could be more meaningful geographical information. Sources like maps could be used to extract where the GPS point is lying, e.g., Is it on a road? Is it in a residential area? Is it in a lake?

In this book, the objective is to extract the semantics (meaningful information) from raw mobile sensing data generated by smartphones, for better understanding the phenomena in the mobile sensing era. In the subsequent chapters, we will select some representative sensors (e.g., GPS, accelerometer) in the mobile phone, and present techniques for identifying semantics of these sensors. For each sensor type, we will design semantic modeling of the sensor data, which can provide different levels of abstraction of sensor semantics; hereinafter we accordingly present and summarize the key supporting techniques to compute the high-level semantics from the low-level raw sensor data stream.

1.2.1 SEMANTIC MODELING OF MOBILE SENSORS

In Section 1.1.1, we summarized a rich list of physical and virtual sensors in the people-centric sensing domain using today's smartphones. These sensors include GPS, WiFi, GSM, accelerometer, gyroscope, barometer, Bluetooth, light, acoustic, etc. In this book, we focus on semantically analyzing three representative sensors that play a significant role in mobile sensing, i.e., *positioning sensors* (like GPS, WiFi) and *motion sensors* (like accelerometer, gyroscope). The fundamental step for extracting semantics from these sensor data is to provide a semantic model to represent sensor streams. In this book, two important modeling techniques will be extensively discussed for modeling sensor data, i.e., "conceptual modeling" and "ontologies."

- **Conceptual Modeling.** Conceptual modeling has been extensively used for formally describing some aspects of the physical and social world around us for the purpose of understanding and communications [16, 116]. It has already played a success role in building conventional information systems. Conceptual models have also been actively used in representing sensor data, e.g., the "stop," "move," and "trajectory" concepts for modeling the abstraction of GPS coordinations [124, 146]. In [152], the authors summarize a set of conceptual models that can be used in sensors in the ubiquitous computing paradigm, particularly focusing on context awareness-oriented applications.

- **Ontologies.** In the Semantic Web, ontologies is the building-block technology for modeling the concepts within a domain, providing "*formal, explicit specification of a shared conceptualization*" [12, 65], etc. Ontologies also plays an important role in modeling sensors and their high-level semantics, e.g., trajectories for GPS data [166], activities for accelerometer data [141], context modeling and reasoning [159], etc.

In line with these conceptual models and ontologies, we present a summary of key semantic modeling approaches for representing sensor data in mobile sensing, and provide formal definitions that could be used for extracting meaningful semantics from raw sensor data. In this book, as we focus on two representative sensor types, the dedicated semantic models will be the following two aspects.

- **Semantic Trajectories.** We can study a sequence of such GPS points of a moving object to infer the set of geographical attributes the moving object is passing on its way. The raw GPS sequence (trajectory) can be represented as a sequence of meaningful *stops* and *moves*, which can be deduced by the velocity of the moving object. Furthermore, the trajectory can be enriched with geographic attributes, e.g., in the morning, the user was moving from home to office, and during the noon time, the user moved from office to a nearby restaurant for lunch. All these are examples of semantics from GPS data, so-called "semantic trajectories" [164, 167].

- **Semantic Activities.** Let us take accelerometer as another example. The raw data of the accelerometer primarily provides the three-axis acceleration components of the device. However, using appropriate machine learning algorithms, scientists have discovered that the readings can be used to reliably infer locomotive states of a user. Examples of such locomotive states are: walk, sit, stand, jump, run, etc. In wearable sensing literature [66, 88], accelerometer worn in multiple body parts can be used to recover fine-grained limb movements. If we take a sequence of such readings over time, a combination of such limb movements can provide us with vital information to infer *complex activities* like making coffee, running on a treadmill, having lunch, etc. These complex activities are often referred to as *activity routines* [13] or high-level *semantic activities* [137, 163] in contrast to the aforementioned locomotive states (also called "micro-activities") such as sit, stand, walk, jog, and drive.

1.2.2 SEMANTIC COMPUTATION FROM MOBILE SENSORS

On one side, this book presents semantic modeling of various sensors in smartphone-based new era of mobile sensing, and particularly focuses on two types of sensors and establishes the dedicated semantic concepts, i.e., "semantic trajectories" and "semantic activities" that were briefly mentioned in the previous section. On the other side, the book provides the reader with an overview of algorithms that are used to progressively enrich a set of sensor data points, coming from smartphone and other embedded sensors, to retrieve these defined semantics corresponding to the data.

In contrast to top-down conceptual and semantic modeling of sensor data representation in the mobile sensing era, we additionally provide extensive summaries of key techniques to compute these semantics from the raw sensor data from a bottom-up viewpoint. Dedicated algorithms about sensor data acquisition, preprocessing (e.g., data cleaning, compression), and statistical modeling will be discussed in details in the book. Many related data mining and machine learning algorithms can be applied into these semantic computations of sensor data. Moreover, high-level semantic annotations, query processing, and visualization of computed semantics are also important facets towards the new era of mobile sensing.

It is worth noting that energy consumption (battery drain) is one of the key issue in smartphone-based mobile sensing. Therefore, we will also provide a dedicated chapter to discuss energy-efficient semantic computation techniques.

We would like to bring up a very important question: Do semantics extracted from mobile sensing reveal private data of users? For example, it has been well studied [20] that the top two significant locations of most users having a city-centric lifestyle reveal their home and their office areas. It is very important that applications and infrastructures designed to "consume" such semantic data preserve privacy of users and provide appropriate tools for users to protect how much personal information flows out and to which stakeholders. Real-world adoption of mobile sensing applications would need to have effective algorithms for extracting semantics as well as first-class privacy preservation algorithms. Thankfully, there has been a significant amount of work in the area of privacy preservation of mobile data. Privacy-preserving data mining [58, 99] investigates how to design new data mining algorithms, while ensuring privacy of users. This book would not cover the topic of privacy, but focus on the knowledge and semantics extraction algorithms from mobile sensing data.

1.3   BOOK STRUCTURE

In the forthcoming chapters, we focus on presenting techniques for knowledge extraction from smartphone sensors, computing semantics of citizen-sensed data, and representation of the sensor semantics at various levels.

In Chapter 2, we present methods of semantic enrichment of location streams captured from GPS sensor of the smartphone, and build so-called semantic trajectories. The key challenge here is to address the inherent unreliability of data sources (e.g., GPS weak signal and noisy data) and to extract various semantic informations (e.g., frequent visit places or mobility behaviors).

In Chapter 3, we discuss methods of recognizing semantic locomotion and activity patterns (named "semantic activities") from acceleration streams, captured from accelerometers of smartphones. Several simple locomotive states (like walk, sit, stand) and complex activities of daily living (including office work, lunch, home cooking, and relax) can be extracted from such motion streams.

In Chapter 4, we address an important challenge of computing semantics from mobile sensors, i.e., the limitation of smartphone battery capacity. A set of state-of-the-art technologies for energy-efficient semantic computation over mobile phone sensors will be summarized. In particular, we will focus on energy-efficient sensing and activity recognition using smartphones.

Finally, Chapter 5 concludes the book and points to other important research and technical challenges in computing semantics in mobile sensing, such as privacy-preserving semantic extraction, and the combination with the emerging platforms of cloud computing as well as mobile social networks.

CHAPTER 2

Semantic Trajectories from Positioning Sensors

Positioning sensors like the GPS form one of the most important embedded sensors in various mobile devices today—an absolutely essential feature for modern smartphones. Berg Insight,[1] for example, forecasts an increase in GPS handsets to one billion units in 2014. As a consequence of this steady growth, the number of applications using positioning data for a variety of purposes is also rapidly increasing. Examples of well-recognized applications of mobility data range from tracking, urban planning, and traffic management, to wildlife movement behavior analysis, mobility-aware social computing, and geo-social network.

Tracking technologies significantly enhance the capabilities of existing applications. They foster new applications and services with location feeds, e.g., ranging from traffic monitoring and environmental management, to land planning and geo-social networks. Here, we list a few examples of application scenarios.

- Scientists implant GPS chips in animals to analyze the gregarious behavior of wild life, e.g., bird migration or monkey habitats in forest.

- Smartphones (e.g., iPhone, Android, Windows phones) can help people establish geo-social networks conveniently and access location-based services, such as Google Now, Foursquare, Facebook Place, Gowalla, and Twitter.

- RFID tags installed in goods can improve the service quality of e-business with better tracking of shipment.

- Vehicles having GPS sensors installed can enhance real-time traffic analysis and provide better planning mechanisms.

We are interested in a set of these applications that typically require tracking of the moving object and storing the sequence of location feeds. These feeds are typically called "trajectories."

In this chapter, we briefly summarize the state-of-the-art for processing positioning/tracking data of humans, vehicles, wildlife, and other objects. We present systematic techniques to extract meaningful information from positioning sensors. We call a trajectory annotated with meaningful information as a "semantic trajectory."

---

[1]http://www.berginsight.com/

2.1   THE POSITIONING SENSOR

In this section, we provide an overview of key positioning techniques and techniques for processing the resulting trajectory data.

2.1.1   SENSOR TYPE AND FUNCTIONALITY

A positioning sensor is a sensing device for estimating the position of a given object. Established and emerging techniques for positioning include GPS, GSM, WiFi/Bluetooth-based localization, and Radio Frequency Identification technology (RFID).

GPS

The Global Positioning System (GPS) is a fully functional navigation system that uses 24 satellites placed in orbit by the U.S. Department of Defense. The satellites circle the Earth and transmit signals to the GPS receivers placed on objects on Earth. By computing the time gap between signal transmission and signal reception, distance between object and satellite can be measured. With various measured distances from multiple satellites, the object position can be estimated. The positioning is presented in terms of the Geodetic Coordinate system, i.e., ⟨*longitude, latitude, altitude*⟩. Additional information can be also computed, such as *speed*, *moving direction*, and *distance to destination*. On average, GPS receivers are accurate to within 15 m. GPS has become one of the standard techniques for tracking mobile devices.

Figure 2.1(a) shows a GPS point located in EPFL campus with coordinates ⟨*longitude = 6.5657 East, latitude = 46.5197 North*⟩. For achieving convenient data analysis and computation, such ⟨*longitude, latitude*⟩ data points need to be converted into the *Cartesian coordinate* system. Different countries typically use their own Cartesian coordinate systems according to a specific time zone, e.g., Swiss coordinate system CH1903[2] in Switzerland. In our experiment, we apply the UTM[3] (*Universal Transverse Mercator*) Cartesian system that work on various datasets. With such transformation, the EPFL ⟨*longitude, latitude*⟩ is transformed to a new point ⟨x, y⟩ = ⟨$313281m, 5154671m$⟩ (see Figure 2.1(b)), where the unit is meter; x is the projected distance of the position eastward from the central meridian, while y is the projected distance of the point north from the equator. Compared to the raw geodetic coordinates, the converted Cartesian coordinates are easier for computation of trajectory features. Examples include calculation of features such as *velocity, acceleration, distance, direction,* and *heading change.* Open-source solutions such as JCoord API[4] are available for supporting the coordinate transformations.

GSM

The GSM (Global System for Mobile communications) is a digital cellular based telecommunication service for transmitting mobile voice and data. GSM network is composed of a set of

---

[2]http://en.wikipedia.org/wiki/Swiss_coordinate_system
[3]http://en.wikipedia.org/wiki/Universal_Transverse_Mercator_coordinate_system
[4]http://www.jstott.me.uk/jcoord/

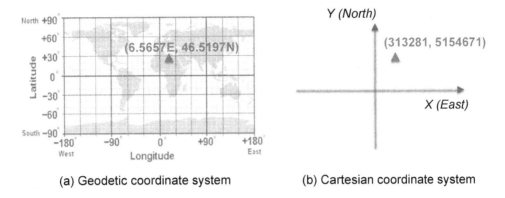

(a) Geodetic coordinate system   (b) Cartesian coordinate system

Figure 2.1: Data transformation of GPS coordinates: $\langle longitude, latitude \rangle \rightarrow \langle x, y \rangle$.

base stations. A combination of stations is responsible for supporting and coordinating communication in a certain area. Each such area is typically called a "cell." By recording the cell log of a mobile device, a sequence of $\langle$cell ID, time$\rangle$ pairs becomes another positioning trace of a device. Typically, well-known triangulation mechanisms[5] are used to find a precise location of the device. The location resolution ranges between 50–200 m and depends on the density of base stations available in the location.

WiFi and Bluetooth

In addition to GPS and GSM, the emerging new generation of WiFi and Bluetooth start to play a new role in positioning. In particular, WiFi and Bluetooth signals can be used for localization in many indoor areas that cannot be captured by GPS or GSM. GPS performs poorly for most indoor environments since the receiver circuitry on the device does not have a direct line-of-sight with the satellites. By taking advantages of increasing wireless access points in urban areas, many companies such as Google and Infsoft have provided positioning service using WiFi. The key idea is to use the signals emitted by the WiFi access points (or Bluetooth transmitters). These signals are received by the device. By modeling the radio propagation or by modeling the resulting radio environment, techniques have been developed to precisely locate a device in an indoor set-up [8, 171].

Recently, with the launch of iBeacons[6] by Apple, the industry has started giving significant attention to Bluetooth-based indoor localization techniques too. The basic structure behind localization remains similar.

For the purposes of this book, we treat different positioning techniques in the same way. Different techniques essentially generate location feeds with varying errors in precisely pointing

---

[5]http://en.wikipedia.org/wiki/Triangulation
[6]http://en.wikipedia.org/wiki/IBeacon

the location. We will use GPS as the representative example for the rest of the chapter discussing techniques of building semantic trajectories.

2.1.2   DEALING WITH POSITIONING DATA: TECHNICAL SUMMARY

In this section, we summarize the key techniques in the literature for processing, managing, and mining GPS positioning data.

Trajectory Data Preprocessing

Positioning technologies invariably introduce errors due to issues such as signal loss, radio model variation, etc. The errors vary as the object moves. Hence, in a trajectory consisting of several positions, the errors are not constant. For GPS based trajectory data, the errors could be due to several reasons. Systematic errors are large distortions due to insufficient visibility of satellites. Ionospheric effects in the upper atmosphere also introduce errors in measurements. Typically, the resulting random errors typically vary within 15 m.

Preprocessing trajectory data is a mandatory step for real-life applications to remove these errors. Certain smoothing and cleaning techniques need to be applied in advance to transform raw mobility data into meaningful records of the trajectory. As time proceeds, the volume of trajectory data can be huge. Hence, compression techniques should be applied in order to reduce the data volume. For network-constrained mobile data (e.g., vehicles driving on the highway), it's useful to map original data points to appropriate road networks. A method called "map matching" is also used for trajectory data cleaning. In this book, we will cover these data cleaning methods for trajectory data preprocessing.

Trajectory Data Management

Database researchers have delved into questions like: How do you efficiently store and retrieve trajectory data? How do you efficiently build index schemes for query processing on top of this data? The core issues are related to building data models, indexing structures, and query processing methods.

Regarding data models, the most popular high-level data model is using continuous functions to approximate a moving object, i.e., $f(time) \rightarrow location$. Therefore, discrete spatiotemporal points (e.g., $\langle x, y, t \rangle$) can be directly used for modeling mobility data; furthermore, approximation functions can be built to store and approximate mobility data. Popular trajectory database prototypes are SECONDO [68], HERMES [126], and PLACE [113]. These are used for efficient storage of the low-level raw mobility data.

The design and construction of an efficient trajectory indexing structure can ensure high performance for efficient querying of the data. The object may move continuously and generate an unending trajectories as time proceeds, resulting in significant storage requirement. Therefore, trajectory indexing is crucial and becomes a possible bottleneck in many trajectory data application systems. One of the first multi-dimensional spatial indexing technique is R-Tree [72]. It is considered as a hierarchical data structure based on B^+-tree in a multi-dimensional space. Many R-tree based extensions have been studied for indexing trajectory data. Two excellent surveys have summarized these trajectory and spatio-temporal indexing techniques [114, 118]. There are also several benchmark systems (e.g., [31, 43, 82, 115]) being established to evaluate these types of indexing methods. These indexing techniques are mostly used for low-level mobility data, i.e., the original raw location records coming from the positioning sensor.

Query processing over trajectory data includes several traditional spatial data querying techniques, such as point & region queries and kNN (k nearest neighborhood) queries. In addition, trajectory and location-based applications require advanced trajectory-based queries, which not only search historical and current trajectory data, but also predict future locations. By adopting the classification of spatio-temporal queries from Pfoser [130] and the moving object database benchmark BerlinMOD [43], we can briefly summarize trajectory queries into two types: *coordinate-based queries* and *trajectory-based queries*. In [155], Theodoridis categorizes ten benchmark queries for the applications of location-based services, among which many are trajectory related queries. These ten queries are divided into four categories: (*a*) *queries on stationary reference objects* including ① point queries, ② range query, ③ distance-based query, ④ nearest-neighbor query, ⑤ topological query; (*b*) *queries on moving reference objects* including ⑥ distance-based query, ⑦ similarity-based query; (*c*) *join queries* including ⑧ distance-join or closest-pair query, ⑨ similarity-join query; and (*d*) *queries with unary operators*, i.e., ⑩ unary operators on spatial and spatiotemporal data type. These queries assist location-based services to access trajectory data. We would not cover trajectory storage and querying further in this book. Readers can find more details in literature such as [56, 67, 174].

Trajectory Data Mining

There are several works that research into new machine learning techniques that can be applied to extract knowledge from the raw trajectory data [62, 75]. The studies typically assume a trajectory as a sequence of spatio-temporal points (x, y, t), where (x, y) is the location coordinate in 2D space and t is the timestamp. Like most conventional data mining studies, trajectory data mining also aims at typical knowledge discovery issues such as *sequential mining, clustering, classification, outliers detection, location prediction*, etc. For example, the mining system MoveMine recently released from Han's group at UIUC [75, 100], includes four main mining functionalities, i.e., (1) periodic pattern mining, (2) swarm and convoy pattern mining, (3) trajectory clustering, and (4) trajectory outlier detection. Basic and advanced concepts of machine learning such as the above form a necessary aspect to extract semantics from trajectory data—the main focus of our chapter. These techniques can be used in the intermediate steps for computing semantic trajectories, for example, to build *stop* and *move* segments and for clustering.

2.1.3   FROM GPS TO SEMANTIC TRAJECTORIES

Extraction of meaningful semantic information from GPS positioning data feeds form an important task. Figure 2.2 briefly shows the idea of computing semantic trajectories from raw GPS feeds. (1) The raw sensor data is a sequence of GPS points. It is hard to understand meaningful information directly from this. (2) The semantic trajectory captures the primary concepts from the data that would be of interest to applications. For example, where the user stays (home, office, market), how she/he moves from one place to another (e.g., using bus, taking metro, or by walking), etc. This chapter discusses technical solutions for generating such semantic trajectories from positioning data.

Figure 2.2:   From GPS sampling points to semantic trajectories.

2.2   SEMANTIC TRAJECTORY MODELING

Traditional trajectory studies largely focus on data management and analysis from a spatial and/or spatiotemporal perspective. Thus, data modeling mainly concerns designing data structures associated with a moving entity to support efficient positioning data management techniques such as indexing and query processing [71, 93].

In order to build a rich mobility data model that can capture high-level semantics like the ones introduced above, conceptual models are being been built to explicitly express the semantics of movement, e.g., the trajectory conceptual view in terms of a stop-move model [146] and trajectory ontologies for conjunctive query processing and reasoning [166]. These modeling concepts are well fitted for the semantic analysis of movements, like tourist movements [5], the semantic interpretation of stops [64], and moves [42]. As a result, such trajectory modeling concepts in the past years have been largely used in several projects on mobility, e.g., GeoPKDD[7] (*Geographic Privacy-aware Knowledge Discovery and Delivery*) [62] and MODAP[8] (*Mobility, Data Mining, and Privacy*). An ontology consists of a set of "concepts" and their associated relationships with each other.

---

[7]http://www.geopkdd.eu/
[8]http://www.modap.org/

In this chapter, two rich trajectory models are discussed, i.e., semantic trajectory ontologies and hybrid semantic model. The ontological infrastructure focuses on the high-level semantic querying and reasoning on trajectories. Due to the time complexity involved in semantic reasoning, typically the data sets are small in this case. The hybrid spatio-semantic model focuses on different levels of data representation, from the low-level raw data generated by positioning devices to the high-level semantic abstraction. Therefore, the choice between these models depends on the application requirements: if the trajectory application is data intensive, for better data abstraction in order to understand basic trajectory semantics, the hybrid model is a good choice; otherwise, if the trajectory application is semantics intensive and needs complex conjunctive trajectory querying and reasoning, the ontological framework can work better.

2.2.1   SEMANTIC TRAJECTORY ONTOLOGY

Next, we describe an ontological infrastructure that addresses the modeling requirements with the goal of supporting creation, management and analysis of trajectory data with a semantic focus. As shown in Figure 2.3, the ontological infrastructure is composed of three ontology modules: *the geometric trajectory module*, *the geography module*, and *the application-domain module*. A module is a sub-ontology of a larger ontology [123]. The benefit of using a modular structure, rather than one all-encompassing ontology, is that it is easier in design and maintenance, coupled with opportunities for query optimization.

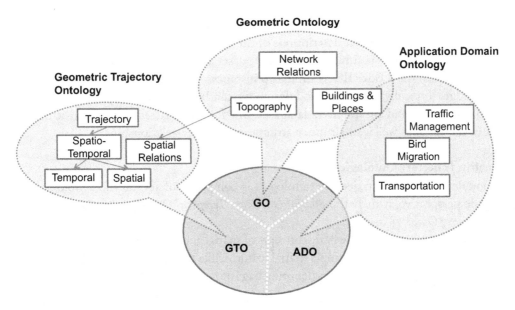

Figure 2.3: Ontological infrastructure for modeling trajectories.

1) *The geometric trajectory module* holds generic concepts for the description of the geometric component of a trajectory. It includes spatio-temporal concepts that are used to specify *spatial*, *temporal* and *spatio-temporal* features needed for a full description of application data. This enables, for example, the spatial and temporal *extents* for describing the trajectory, e.g., the spatio-temporal lines (a sequence of $\langle x, y, t \rangle$ points) formed by moving points during a trajectory.

2) *The geography module* captures the concepts that describe the properties of the underlying terrain. Concepts include those describing the topography of the land, natural (e.g., hydrological) and artificial (e.g., routes, trains, facilities) networks, buildings, landmarks, vegetation, and anything else that is related to the underlying trajectory data and of interest to the application. This module is tightly coupled with the geometric trajectory module because applications require such geographic information to link the mobility data with the corresponding spatial extent—e.g., road network information and point of interests such as buildings and semantic places.

3) The *application domain* module gathers all application-dependent/domain-specific concepts. Examples of such modules include traffic management ontologies (including concepts like traffic jam, traffic light, crossing), bird migration ontologies, transportation ontologies, and tourism ontologies. These ontologies can provide more meaningful knowledge and dedicated concepts of certain applications.

Ontological knowledge is usually represented using two basic concepts—TBox and ABox [148]. The TBox holds abstract descriptions of the concepts, their properties (i.e., data type) and the links (i.e., roles). This is similar to defining classes and properties. The ABox holds instances of the elements in the TBox. This is similar to instances of a class. Applications need both levels of knowledge. Typically, the three modules above can be modeled using combination of TBox and ABox concepts and are stored in databases. The following subsections describe the details of the three modules, as well as their inner-structures and inter-relationships.

Geometric Trajectory Ontologies

The module of geometric trajectory ontologies is composed of several sub ontologies from the geometric perspective, namely the *Spatial Ontology*, *Temporal Ontology*, *Spatiotemporal Ontology*, and *Trajectory Ontology* itself.

Figure 2.4 summarizes the basic concepts of spatial and temporal ontologies. The *SimpleGeo* concept generalizes any kind of simple spatial features, like *points*, *lines*, and *surfaces*. The *Complex-Geo* concept generalizes the concepts denoting bags of points, lines, and surfaces, and also denotes any heterogeneous complex extent. The *SimpleTime* and *ComplexTime* concepts are defined in a similar fashion. The concept terminologies in the figure are adopted from the MADS (Modeling of Application Data with Spatio-temporal features) specifications [124]. Towards the modeling of "space," similar but not identical concepts have been proposed by the Open Geospatial Consor-

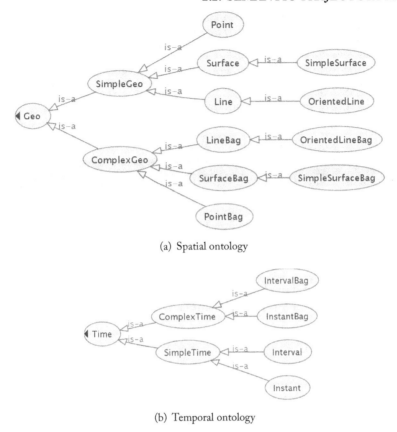

(a) Spatial ontology

(b) Temporal ontology

Figure 2.4: Example concept hierarchies in spatial and temporal ontologies.

tium (OGC[9]) and ISO Technical Committee 211 (Geographic Information/Geomatics).[10] These concepts are expected to become universal standards and whenever that happens this dedicated ontology needs to be compatible, as concepts may be replaced by standard data types. The same is on its way for the temporal domain (see e.g., [120]).

The *Spatiotemporal Ontology* builds on the spatial and temporal ontologies to provide concepts for describing phenomena with time-varying geometry. For example, to track the movement of a car, the car geometry (i.e., its spatial extent) is described using the concept of time-varying point, i.e., a point whose $\langle x, y \rangle$ position changes over time. While movement is continuous in real life, its data record usually consists in a sequence of discretized timestamped points (i.e., $\langle x, y, t \rangle$); furthermore, certain interpolation functions are needed to compute the $\langle x, y \rangle$ position given any time t' not necessarily stored in a $\langle x, y, t \rangle$ triple. Adopting such interpolation aspects,

[9]http://www.opengeospatial.org/
[10]http://www.isotc211.org/

the time-varying geometry can be defined in description logic (DL) as a concept with at least one instance of the *listOf* role that links it to *TimePoint* instances (representing the $\langle x, y, t \rangle$ triples). The same goes for *moving lines*, *moving surfaces*, and *moving objects* of undefined spatial type. See Table 2.1 for the formal definitions. For a full and rigorous discussion about the definition of spatio-temporal concepts the reader can refer to the foundational work of Güting on moving object databases [49] [67] and recent work on trajectory databases [93].

Table 2.1: Definition of spatiotemporal ontology

Concept	Definition in Description Logic
TimePoint	$\sqsubseteq \exists$ hasGeometry.Point $\sqcap \exists$ hasTime.Instant
TimeVaryingPoint	$\sqsubseteq \exists$ listOf.TimePoint
TimeLine	$\sqsubseteq \exists$ hasGeometry.Line $\sqcap \exists$ hasTime.Instant
TimeVaryingLine	$\sqsubseteq \exists$ listOf.TimeLine
TimeSurface	$\sqsubseteq \exists$ hasGeometry.Surface $\sqcap \exists$ hasTime.Instant
TimeVaryingSurface	$\sqsubseteq \exists$ listOf.TimeSurface
TimeGeo	$\equiv$ TimePoint $\sqcup$ TimeLine $\sqcup$ TimeSurface
TimeVaryingGeo	$\equiv$ TimeVaryingPoint $\sqcup$ TimeVaryingLine $\sqcup$ TimeVaryingSurface

Finally, the *Trajectory Ontology* holds the concepts that describe how movement can be understood as a set of structured trajectories. The needed concepts have been identified and discussed in [146]. They include the *stop* and *move* concepts. They define a segmentation of the trajectory. They also include the *Begin* and *End* concepts denoting the $\langle x, y, t \rangle$ triple that denotes where and when the trajectory starts and the $\langle x, y, t \rangle$ triple indicating the end of the trajectory. This is also referred to as the "Begin-End-Stop (B.E.S)" concept set. The B.E.S. set of concepts allows us to define a design pattern for structured trajectories and supports linking trajectory elements to application elements, thus allowing semantic enrichment on the trajectory data to provide better mobility understanding.

Table 2.2 summarizes the formal definition of trajectory and its related concepts. These definitions relate to moving point trajectories and consider a point type geometry for *begin*, *end*, and *stops*. From the temporal perspective, *begin* and *end* are associated with a time instant, while *stops*, which may have a temporal duration, are associated with a time interval. *Moves* are sections of a trajectory where there are no *stops*. *Moves* connect one *stop* to the next *stop*, with the exception of the first *move* that starts at *Begin* and the last *move* that ends at *End*. Each *trajectory* has exactly one *Begin*, exactly one *End*, one or more *Move*, and zero or more *Stop* instances.

Geography Ontology

This ontology concentrates on concepts of the geographical environment, and these concepts are generic and application dependent [109][140]. Existing specifications and standards can be reused to build this ontology. A few dedicated associations such as the *World Wide Web Consor-*

Table 2.2: Definition of trajectory ontology

Concept	Definition in Description Logic
Begin	$\sqsubseteq$ TimePoint
End	$\sqsubseteq$ TimePoint
Stop	$\sqsubseteq \exists$ hasGeometry.Point $\sqcap \exists$ hasTime.Interval
B.E.S	$\equiv$ Begin $\sqcup$ End $\sqcup$ Stop
Move	$\sqsubseteq$ TimeVaryingPoint $\sqcap \exists$ from.B.E.S $\sqcap \exists$ to.B.E.S
Trajectory	$\equiv \exists$ hasBegin.Begin $\sqcap$ =1 hasBegin $\sqcap \exists$ hasEnd.End $\sqcap$ =1 hasEnd $\sqcap \exists$ hasMove.Move $\sqcap \forall$ hasStop.Stop

tium (W3C)[11] and the *Ordnance Survey* (OS),[12] have devised relevant ontologies for describing geographical concepts based on the Open Geospatial Consortium (OGC) standards. For example, a domain ontology *BuildingsAndPlaces* contains concepts and properties describing human constructions in geographical space, such as *banks*, *churches*, and *bowling clubs*. Typical properties in this context are *hasName*, *hasHistoricInterest*, *hasFootprint*, etc. Rather than designing a new geography ontology from scratch, importing an existing ontology or a portion of it can efficiently provide a first version; and then the initial version can be manually augmented with extra concepts while deleting concepts that are irrelevant to the application at hand.

Application Domain Ontology

An application domain ontology holds the wide range of concepts that make up the *universe of discourse* for a set of applications in the same domain. Domain ontologies have been intensively investigated in the literature. Examples in the traffic management domain include the *Ontologies of Transportation Networks* (OTN) [103], *Towntology*,[13] and *Spatially Aware Information Retrieval on the Internet* (SPIRIT).[14] Some of these approaches do not make an explicit distinction between the *Geography Ontology* and the *Traffic Management Ontology*, but focus on one of them or combine the two. While the ultimate result may be the same, the initial distinction of concepts provided by a modular approach promotes reusability and understandability. For developing ontologies for real-life deployment, we can assume that the geography and application domain ontologies may overlap in some applications. Geography knowledge can be extracted from many free map data sources, e.g., Openstreetmap. Additional application-domain knowledge can be used to build concepts like *office* and *home* by utilizing some additional information sources.

Table 2.3 illustrates some axioms for a traffic management application. The first axiom defines a *HomeOfficeTrajectory* as a trajectory that starts at home and ends in office. The third axiom defines a *BlockedStreetT* as a street where there is some long-term road work ongoing. The

[11]http://www.w3.org/2005/Incubator/geo/XGR-geo-ont-20071023/#crs/
[12]http://www.ordnancesurvey.co.uk/oswebsite/ontology/
[13]http://www.towntology.net/
[14]http://www.geo-spirit.org/

fourth axiom describes the concept of a high congestion trajectory as a trajectory that has more than ten stops with a defined long time interval.

Table 2.3: Definition of traffic management ontology

Concept	Definition in Description Logic
HomeOfficeTrajectory	$\equiv$ Trajectory $\sqcap$ $\exists$hasBegin.$\exists$isLocatedIn.Home $\sqcap$ $\exists$hasEnd.$\exists$isLocatedIn.Work
LongTermRoadWork	$\sqsubseteq$ RoadWork
BlockedStreetT	$\equiv$ StreetT $\sqcap$ $\exists$ hasRoadWork.LongTermRoadWork
HighCongestionTrajectory	$\equiv$ Trajectory $\sqcap$ $\exists_{>10}$ hasStop. $\exists$ hasTime.LongTimeInterval
LongTimeInterval	$\sqsubseteq$ Interval

The Complete Modular Ontology

Combining the *Geometric Trajectory Ontology*, *Geography Ontology* and *Traffic Management Ontology* ontologies together leads to the final overall *Semantic Trajectory Ontology*. This final ontology provides the full semantic description of application-relevant trajectories with their domain specific semantic meaning. Figure 2.5 briefly depicts the final semantic trajectory ontology for a traffic management application. For the sake of clarity, the figure shows only a part of this ontology with the most important concepts and relationships.

2.2.2   HYBRID SEMANTIC TRAJECTORY MODEL

Once we have defined the trajectory ontology concepts, an important question arises: *How do we populate these concepts from low-level mobility data feeds like GPS tracking points?* Mobility feeds are essentially a time series of location coordinates where the moving object has passed. Given the heterogeneity of different semantic concepts, how do we organize them so that we can have a systematic means of developing these abstract concepts from low-level data feeds? We call this a *hybrid semantic trajectory model*. With this hybrid model, we can easily conceptualize a supporting platform that can progressively populate these concepts from raw data feeds. The hybrid semantic trajectory model should support multi-level trajectory abstractions, ranging from the raw mobility data to high-level semantic trajectories. The key design considerations for this hybrid model are as follows.

- **Raw Data characteristics.** The model should consider characteristics of raw mobility tracking data (e.g., spatial and temporal gaps, uncertainties) to create simple *low-level* representations (e.g., hourly, daily, monthly and geo-fenced trajectories).

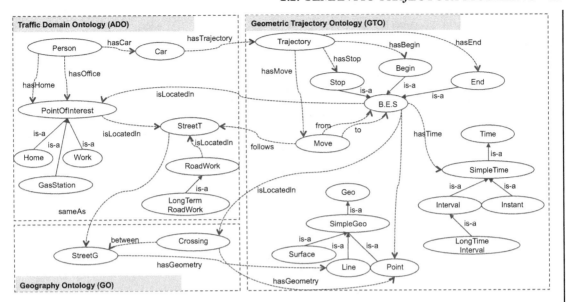

Figure 2.5: Semantic trajectory ontology for a traffic management application.

- **Progressive computation.** The model should be designed so that a layered computing platform can generate higher-level semantic abstractions from the underlying lower-level trajectory representation.

- **Encapsulate various semantics.** The model should be able to encapsulate various kinds of semantic annotations inferred from heterogeneous third-party geographic artifacts (e.g., landuse, road networks, points of interests) and relevant rules/regulations pertaining to the real world (e.g., cars stop at red lights, buses stop at bus stops).

Therefore, a hybrid model may be considered as one which consists of (1) the *Raw Data Model* that provides the trajectory definitions available from the raw data perspective directly collected from mobile devices; (2) the *Conceptual Model* that is a mid-level abstraction of a trajectory that provides a structured view of the raw mobility data; and (3) the high-level *Semantic Model* that provides a semantically enriched and more abstract view of the trajectory. Figure 2.6 provides an illustration of these models.

Raw Data Model

The raw data model is the first abstraction level over the raw mobility data. The raw data, like GPS records, are typically captured by positioning sensors that continuously record the location of the moving object. So, the raw mobility data for a moving object is in essence a long sequence of spatio-temporal tuples (*position, timestamp*) collected over some time interval. Most real-life

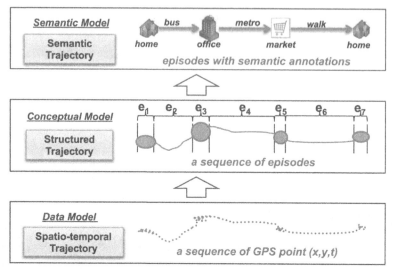

Figure 2.6: The hybrid spatio-semantic trajectory model.

location traces today are essentially GPS-like tuples (*longitude, latitude, timestamp*) − (x, y, t) in short. From now on, the term *GPS feed* shall be used to represent the *raw* sequence of spatio-temporal points of a moving object.

In the *Raw Data Model*, each GPS feed is decomposed into *subsequences* so that each sub-sequence represents one meaningful unit of movement. These meaningful units are called "*spatio-temporal trajectories*". Consequently, a spatio-temporal trajectory has a starting point (x, y, t), called *Begin*, and, further, an ending point, called *End*; These two spatio-temporal points delimit the subsequence/extent of the trajectory, along with the corresponding time interval $[t_{begin}, t_{end}]$.

Definition 2.1 (Spatio-temporal Trajectory – $\mathcal{T}_{spa}$)   Given a GPS feed $\mathcal{G}$ of a moving object, $\mathcal{G} = \{p_1, p_2, \ldots, p_m\}$ (where each $p_i = (x_i, y_i, t_i)$ represents a spatio-temporal point), a spatio-temporal trajectory $\mathcal{T}_{spa}$ is a result of $\mathcal{G}$ for a given time interval $[t_{begin}, t_{end}]$, such that the subsequence does not contain any significant space or time gap.

Conceptual Model

The term conceptual model refers to the logical partitioning of a spatio-temporal trajectory $\mathcal{T}_{spa}$ into a series of non-overlapping *episodes*. A $\mathcal{T}_{spa}$ partitioned into episodes is called a *Structured Trajectory* ($\mathcal{T}_{str}$). Conceptually, an *episode* abstracts a subsequence of spatio-temporal points in $\mathcal{T}_{spa}$ that show a high degree of correlation w.r.t. some spatio-temporal feature (e.g., velocity, angle of movement, density, time interval). An *episode* has the following salient features:

- **It is a generic trajectory structuring concept.** By generically denoting a subsequence of a trajectory, the episode concept generalizes several other concepts that have been defined in the literature. For example, *stop* and *move* episodes were defined in [146]. In [6], the authors visualize trajectories as sequences of time-bars that can be considered as episodes defined according to range intervals of a given attribute (e.g., distance to a given geo-object, speed, direction).

- **It can be computed automatically.** Episodes can be automatically computed with relevant trajectory structuring algorithms by using the correlations in the spatio-temporal characteristics of consecutive points of the GPS feed, e.g., neighboring points with the same *velocity* are merged together as an episode. Similarly, other attributes like *acceleration, orientation,* and *density* can be also used for such automatic computation.

- **It enables data compression.** Instead of adding semantics to each GPS record (which is possible), we can tag a whole episode, which consists of multiple GPS records. This reduces the size of the data needed to represent structured trajectories. For instance, Figure 2.6 shows the annotation of seven episodes in the conceptual model ("S" and "M" annotations), which is more efficient than annotating the original hundreds of GPS points in the data model.

Definition 2.2   (Structured Trajectory – $\mathcal{T}_{str}$)   A structured trajectory $\mathcal{T}_{str}$ consists of a sequence of "episodes," i.e., $\mathcal{T}_{str} = \{e_1, e_2, \ldots, e_m\}$, where each episode $e_i = (time_{from}, time_{to}, da, rep)$:

- "$time_{from}$" is the instant of the first point of the episode, $time_{to}$ is for the last point of the episode;

- "da" is the "defining annotation" of the episode. It represents the common spatio-temporal characteristic that is shared by all the spatio-temporal points inside the episode—which could be the same location, the same velocity, etc.; and

- "rep" is the spatio-temporal or spatial representation of the episode. It is either the sequence of points of the episode or a spatial abstraction of this sequence, e.g., the two extremity points of the episode, the center point of the episode, or the bounding rectangle of the episode.

Semantic Model

In the *Semantic Model*, a semantic trajectory $\mathcal{T}_{sem}$ is a structured trajectory enhanced with semantic annotations of its episodes. An example of *semantic trajectory* is shown in the top layer of Figure 2.6. It shows the semantic trajectory of a given employee on a given day: she/he goes to work from *home* in the morning; after finishing the *work* in later afternoon, he leaves for shopping in *market*, and finally reaches *home* in the evening.

Semantic trajectories can be computed by integrating data from third-party geographic sources (e.g., geographic databases describing landuse, road network, or points of interest), social networks containing data related to locations, and common sense knowledge about the real world. For example, usually GPS points collected around midnight are located at home, provided the points do not demonstrate a lot of mobility. In the subsequent section, we describe methods to semantically enrich trajectories using such third-party data sources. For now, we can define a Semantic Trajectory.

Definition 2.3 (Semantic Trajectory $\mathcal{T}_{sem}$)   A semantic trajectory $\mathcal{T}_{sem}$ is a structured trajectory where the spatial data (the coordinates) are replaced by geo-annotations and further semantic annotations may be added. Episodes are enriched to generate semantic episodes *(noted as "se")* with geographic or application knowledge: the spatio-temporal or spatial representation of the episode is replaced by a reference to the geo-object where the episode takes place, i.e., $\mathcal{T}_{sem} = \{se_1, se_2, \ldots, se_m\}$, where each semantic episode is defined by: $se_i = (da, sp_i, t_{in}^{(sp_i)}, t_{out}^{(sp_i)}, tagList)$:

- "*da*" is the defining annotation of the episode (e.g., "*stop*" or "*move*");

- "*sp_i*" *(semantic position)* is a geo-object. The geo-object represents the location of the episode at the semantic level. It is a real-world object taken from the available geographic knowledge (e.g., a building, roadSegment, administrativeRegion, landuse region) or from application domain knowledge (e.g., the home or the office of a specific person of the application). A frequent characteristic of geo-objects used for semantically locating episodes is the type of the geo-object, e.g., Hotel, Restaurant, LocalStreet, CollectorStreet;

- "$t_{in}^{(sp_i)}$" is the incoming timestamp for the trajectory entering this semantic position (sp_i), and $t_{out}^{(sp_i)}$ is the outgoing timestamp for the trajectory leaving sp_i. They can be approximated by the *time$_{from}$* and *time$_{to}$* of the episode; and

- "*tagList*" is a list of additional semantic annotations about the episode, e.g., the activity performed during stop episodes by the moving object (shopping, working or eating), the transportation mode used by the moving object for the move episodes (bike, bus, car, or walk).

2.3   SEMANTIC TRAJECTORY COMPUTATION

The previous section was dedicated towards discussing different ontological and semantic models for trajectory representation. In this section, we describe methods to use these models for populating semantic trajectories from low-level GPS feeds. We organize the methods in the form of a framework that we call *Trajectory Computing Platform*. We describe the different layers in the platform and algorithms involved in each layer. The *Trajectory Computing Platform* exploits the trajectory model and builds trajectory instances at different levels (*spatio-temporal*, *structural*), from large-scale real-life GPS feeds. Figure 2.7 shows the three layers of the platform, each containing several techniques for progressive computation of the trajectory instances.

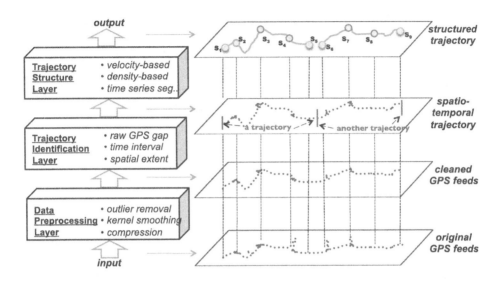

Figure 2.7: Trajectory computing platform.

The *Data Preprocessing Layer* is the lowest layer. It cleans the raw GPS feed, in terms of preliminary tasks such as removal of outliers and regression-based smoothing. The outcome of this step is a cleaned sequence of (x, y, t). The *Trajectory Identification Layer* is the middle layer. divides the sequence of cleaned (x, y, t) points into several meaningful trajectories (spatio-temporal trajectories $\mathcal{T}_{spa}$). This step exploits gaps present in the sequence and applies well-defined policies for temporal and spatial demarcations (e.g., daily time intervals, city areas, etc.). The *Trajectory Structure Layer* is the top-most layer and computes *episodes* present in each spatio-temporal trajectory and generates structured trajectory $\mathcal{T}_{str}$. It contains several algorithms for computing correlations between consecutive GPS points.

2.3.1   DATA PREPROCESSING LAYER

Due to GPS measurements and sampling errors from mobile devices, the recorded position of a moving object is not always correct [172]. Usually the recorded data is unreliable, imprecise, incorrect, and contains noise. There is work involved in determining possible causes for such uncertainty [56]. A *Data Preprocessing Layer* is used for several low-level actions on the raw GPS data that lead to a cleaner and concise representation of the GPS feed. For example, this layer can be used to detect *systematic errors* (or called "outliers"). Outliers are observations that deviate significantly from the desired correct position. For detecting outliers, velocity threshold can be applied to remove points that do not give a reasonable correlation with expected velocity, within a time window of readings. Each GPS feed has domain knowledge of the moving object (e.g., car, bike, people walk). This allows removal of outliers by using the maximum velocity of this kind of moving entity.

Apart from outliers, the data also contains noise. GPS signals can have noise from several sources. For example, ionospheric effects and clocks of satellites can contribute towards white noise of ± 15 m.[15] We describe a Gaussian kernel based local regression model to smooth out the GPS feed. In this method, the smoothed position $(\widehat{x_{t_i}}, \widehat{y_{t_i}})$ at time t_i is the weighted local regression based on the past points and future points around time t_i – all points at t_j within a sliding time window - σ, i.e., $|t_i - - - t_j| \leq \sigma$. Formula 2.1 presents the detailed smoothing function:

$$(\widehat{x_{t_i}}, \widehat{y_{t_i}}) = \frac{\sum_j k(t_j)(x_{t_j}, y_{t_j})}{\sum_j k(t_j)}, \tag{2.1}$$

where $k(t_j) = e^{-\frac{(t_j - t_i)^2}{2\sigma^2}}$ is the Gaussian kernel function that defines the smoothing window size as the kernel bandwidth and controls the width of the Gaussian bell. To control the smoothing related information loss, a reasonably small value for σ (e.g., 5 $\times$ GPS sampling frequency) is adopted so that only nearby points from temporal perspective can affect the smoothed position. This is necessary to calibrate the technique to handle only the noise while avoiding under-fitting.

Figures 2.8, 2.9, and 2.10 show an example of the smoothing algorithm on a real data set taken from wildlife tracking data on a given day. It contains 52 GPS (x,y,t) records. Figure 2.8 shows the smoothed longitude (actually transformed X in Cartesian coordinate). Figure 2.9 shows the smoothed latitude (transformed Y in Cartesian coordinate) and Figure 2.10 plots the original GPS feed before and after smoothing.

These smoothing techniques are designed for GPS feeds of objects that may move without any constraints, e.g., people trajectories, animal foraging movements, etc. However, in many cases, objects (e.g., vehicles) move along defined network constrained paths (e.g., transportation network) [69]. Researchers have designed map matching algorithms [18] for such network-constrained trajectory data. In map matching, GPS feeds (or trajectories) are "matched" with maps of spatial road networks. This can be used to determine several pieces of information like the correct road segment sequences on which a vehicle is moving [134], annotating trajectories, in particular the *move* episodes. This can also be used to remove noise and outliers. Section 2.4.2 will provide some details of a candidate map-matching algorithm.

Apart from outlier detection and noise removal, the data preprocessing layer may also have algorithms to compress the data for a concise representation of the GPS feed [56]. We briefly

---

[15]www.kowoma.de/en/gps/errors.htm

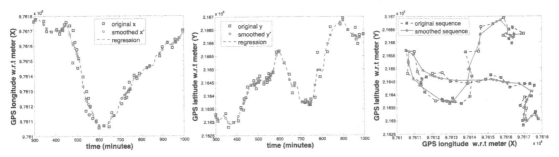

Figure 2.8: Smooth GPS (x).   **Figure 2.9:** Smooth GPS (y).   **Figure 2.10:** Original/smoothed.

describe two representative trajectory data compression algorithms in this chapter. The first one is called the Douglas-Peucker extensions with the application of Synchronized Euclidian Distance (SED) [110]. The second method is called STTrace. It contains trajectory summarization metrics that are computed by using information like speed and direction [131].

Synchronized Euclidean Distance (SED):

SED is an error metric that projects each of the points in a sequence on a straight line drawn from the first to the last point in the sequence (see Figure 2.11). In this example, let P_b be the currently examined point against line $P_1 P_n$, we have a projected point P'_b using a uniform velocity projection, and the SED between P_b and $P_1 P_n$ is the Euclidian distance between P_b and P'_b. Formula 2.2 provides the detailed SED computation.

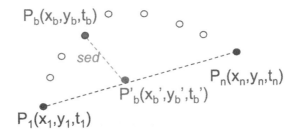

Figure 2.11: Synchronized Euclidean distance.

$$
\begin{aligned}
x'_b &= x_1 + (x_n --- x_1) \times \frac{t_b - t_1}{t_n ---- t_1} \\
y'_b &= y_1 + (y_n --- y_1) \times \frac{t_b - t_1}{t_n ---- t_1} \\
sed(P_1, P_b, P_n) &= \sqrt{(x_b --- x'_b)^2 + (y_b --- y'_b)^2}
\end{aligned}
\tag{2.2}
$$

The main idea of using SED for data compression is to recursively compute the SED distances for each point within a sequence, and compare it with a given threshold ε to judge whether the point can be removed or not. Figure 2.12 provides an example. In Figure 2.12(a), the SED values of all the points in the middle are calculated (i.e., $sed(P_1, P_i, P_7)$ for all $i \in [2, 6]$). We choose the *most representative* data point (P_4) as the one having the largest SED values compared with other points (i.e., P_2, P_3, P_5, P_6). Therefore, the algorithm keeps P_4 and recursively goes to the next loop for subsequence $P_1 P_4$ and $P_4 P_7$. Figure 2.12(b) shows the later steps for computing the new SED values for the intermediate points in segment $P_1 P_4$ and $P_4 P_7$, i.e., $sed(P_1, P_2, P_4), sed(P_1, P_3, P_4)$ in $P_1 P_4$ & $sed(P_4, P_5, P_7), sed(P_4, P_5, P_6)$ in $P_4 P_7$. It discovers that all of these have $sed < \varepsilon$. Therefore, all of these four intermediate points are removed (P_2, P_3, P_5, P_6) and the new compressed trajectory is just $P_1 \rightarrow P_4 \rightarrow P_7$. The final compressed trajectory data has only three spatio-temporal points instead of the original one with 7 points, achieving the compression ratio of 3/7, approximately 42.86%.

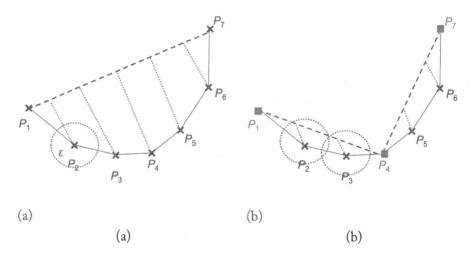

(a)   (b)

Figure 2.12: DP extensions for data compression using SED.

STTrace: This method applies the concept of a *safe area* to judge whether the points can be removed or not. This safe area (or sector) is calculated by the moving object's *velocity* and *direction*. For example, in Figure 2.13(a), based on the $speed_{min}$, $speed_{max}$ and previous direction $P_1 P_2$, a *sector* area may be computed for future points P_3, P_4. We find that sector $P_3 \in S_3$ while $P_4 \notin S_4$. Therefore P_3 can be removed whilst P_4 needs to be kept. Similarly, in Figure 2.13(b), the algorithm can remove P_5 but needs to keep P_6. Finally, we achieve the compressed trajectory data as $P_1 \rightarrow P_2 \rightarrow P_4 \rightarrow P_6 \rightarrow P_7$.

The SED based method is an offline method as it requires knowledge of the whole sequence. The second method (STTrace) can be applied online to an incoming stream since safe sectors can be computed based on real-time parameters like velocity and direction.

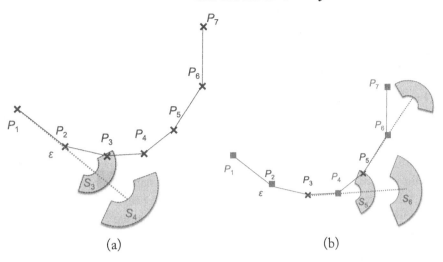

Figure 2.13: Direction- and speed-based STTrace for data compression (Left: Determine P_3, P_4; Right: Determine P_5, P_6, P_7).

2.3.2   TRAJECTORY IDENTIFICATION LAYER

This layer uses the cleaned data and extracts relevant non-overlapping spatio-temporal trajectories $\mathcal{T}_{spa}$ (*data model*). The central issue here is to determine reasonable identification policies, to identify the *division points* (x_i, y_i, t_i) that divide the continuous GPS feeds into consecutive trajectories at appropriate positions. We present a few identification policies for various trajectory scenarios.

Policy 1 (Raw GPS Gap) *Divide the sequence of (x, y, t) GPS records into several spatio-temporal trajectories according to the GPS gaps that satisfy one of the following conditions.*

1. *Given a large time interval $\Delta_{duration-large}$, if two consecutive GPS records, $p_i(x_i, y_i, t_i)$ and $p_{i+1}(x_{i+1}, y_{i+1}, t_{i+1})$, are such that the temporal gap $t_{i+1} - t_i > \Delta_{duration-large}$, then p_i is the ending point of the current trajectory whilst p_{i+1} is the starting point of the next trajectory.*

2. *Given both a time interval $\Delta_{duration}$ and a spatial distance $\Delta_{distance}$, if two consecutive GPS records, p_i and p_{i+1}, are such that the temporal gap $t_{i+1} - t_i > \Delta_{duration}$ and the spatial gap $\sqrt{(x_{i+1} - x_i)^2 + (y_{i+1} - y_i)^2} > \Delta_{distance}$, then p_i is the ending point of the current trajectory whilst p_{i+1} is the starting point of the next trajectory.*

This policy utilizes the significant temporal (and spatial) gaps in the GPS feed for separating two consecutive spatio-temporal trajectories $\mathcal{T}_{spa}$. GPS trajectories often exhibit such gaps due to several reasons. For example, tracking devices usually turn off the GPS if the object does not move for a long while (to save power) or if there is no satellite coverage (indoor locations). The first sub-policy exploits large temporal gaps $\Delta_{duration-large}$ to extract $\mathcal{T}_{spa}$. This is typically relevant for vehicle movement scenarios. The second sub-policy uses both temporal and spatial gaps, where the two parameters are determined by statistical analysis of GPS feeds (e.g., gap distribution, type of movement: vehicular, pedestrian, etc.).

Policy 2 (Predefined Time Interval) *Divide the stream of GPS feed into several subsequences contained in given time intervals, e.g., hourly trajectory, daily trajectory, weekly trajectory, and monthly trajectory.*

This policy allows meaningful division of a GPS feed into periods for analyzing mobility behaviors. Short-term period is particularly relevant for human movements (e.g., daily movement of weekday behavior analysis). Wildlife monitoring on the other hand needs to capture longer-term trajectory behaviors such as monthly or seasonal patterns (e.g., yearly movement analysis for the bird migration scenario).

Policy 3 (Predefined Space Extent) *Divide the stream of GPS feed into several subsequences according to a spatial criteria, e.g., fixed distance, geo-fenced regions, and movement between predefined points for network constrained trajectories.*

This policy allows the division of a GPS feed according to the covered distance (e.g., every 20 miles); according to a specific area (e.g., trajectories in an university campus), where trajectories are defined when the object enters or exits the area; or between two given positions.

Policy 4 (Time Series Segmentation) *Divide the stream of GPS feed into several subsequences according to a (semi-) automatic algorithm for segmenting time series, based on spatial or/and temporal correlations.*

Trajectory data in essence is a special kind of time series, where the values are the locations $\langle x, y \rangle$ as time flows. Therefore, conventional time series segmentation algorithms can be applied for trajectory identification. Keogh et al. [86] categorizes time series segmentation methods into three types: sliding window, top-down, and bottom-up. These methods are used for time series based segmentation of the mobility data. Policy 2 and Policy 3 can be considered as sliding window-based methods, where the window is dynamically determined by the given temporal intervals or spatial areas. The top-down and bottom-up methods can generate excessive fragmentation of trajectories (i.e., a lot of small segments), which is not good for the *trajectory identification* step. Nevertheless, they can be applied for the *trajectory structuring* step.

The choice of the trajectory identification policy (from Policy 1 to Policy 4) depends on the application and data characteristics (e.g., with or without big gaps). For example, people trajectories created by users carrying smartphones may use Policy 2 in order to come up with diurnal trajectories. On the other hand, trajectories created by taxi cabs can be divided according to city zones by using Policy 3, or by dividing the trajectories into intra-city and trajectories that fork outside of the city boundaries.

2.3.3   TRAJECTORY STRUCTURE LAYER

After identifying separate spatio-temporal trajectories, the next task is to compute their internal structures, constructing structured trajectories $\mathcal{T}_{str}$ that consist of meaningful episodes. The core issue in this *trajectory structure* layer is to group consecutive GPS points into an episode. Researchers have worked on creating algorithms for *velocity*, *density*, *orientation*, and *time series* based identification of episodes. There are two kinds of episodes—*stop* and *move*—that form a common requirement across many trajectory applications. Hence, we focus on how to detect these episodes. Next, we present two representative methods for computing trajectory structure, i.e., *velocity-based* and *density-based* structure identification methods.

Velocity-based Trajectory Structure Discovery

The idea is to determine whether a GPS point $p(x, y, t)$ belongs to a stop episode or a move episode by using a speed threshold (Δ_{speed}). A simple implementation is: *if the instant speed of p is lower than Δ_{speed}, it is a part of a stop, otherwise it belongs to a move.* Figure 2.14 traces the speed evolution of a vehicle, showing how stops can be determined by a given Δ_{speed}. Besides Δ_{speed}, we also use a second parameter—*minimal stop time τ*—in order to avoid false positives. False positives can be generated, in particular for trajectories having short-term *congestions* with a low velocity. In this case, a temporary congestion may be mistaken as a stop.

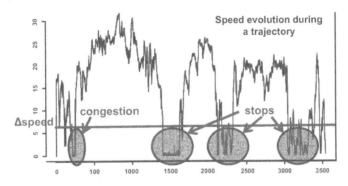

Figure 2.14: Velocity-based stop identification.

Determining a suitable value for Δ_{speed} is a challenge: *if Δ_{speed} is too high, many stops appear; on the contrary, if Δ_{speed} is too low, probably no stops are computed.* Figure 2.14 simply shows a constant Δ_{speed} applied all across the trajectory. However, this is not practical in real-world scenarios, where the value of Δ_{speed} should rather be flexible according to the context of the moving object. For example, vehicles with different levels of performance like bicycles or motor cars have different operating speed characteristics. Similarly, different road networks (is it a highway or a secondary road path?), and different weather conditions (sunny or snowy days) call for diverse speed thresholds. Although it is possible to get this contextual information, it would substantially increase the number of information sources that need to be integrated. To circumvent this, a different approach has been researched where a generic method is designed for determining Δ_{speed}, based on the classes of moving objects in the area of consideration and their aggregate statistics. To describe this, we first define a Dynamic Velocity Threshold.

Definition 2.4   (Dynamic Velocity Threshold - Δ_{speed})   For each GPS point $Q(x, y, t)$ of a given moving object (obj_{id}), the Δ_{speed} is dynamically determined by the moving object (by using $\overline{objectAvgSpeed}$—the average speed of this moving object) and the underlying context (by $\overline{positionAvgSpeed}$—the average speed of most moving objects in this position $\langle x, y \rangle$), i.e., $\Delta_{speed} = min\{\delta_1 \times \overline{objectAvgSpeed}, \delta_2 \times \overline{positionAvgSpeed}\}$, where δ_1 and δ_2 are coefficients.

In this definition, $\overline{objectAvgSpeed}$ is easy to calculate. This is the average speed of the moving object. The $\overline{positionAvgSpeed}$ is the average speed of the moving objects in the position $\langle x, y \rangle$. We need to approximate it by using space division. To do this, we divide space into regular cells and the average speed is calculated in each cell $\overline{cellAvgSpeed}$. For network-constrained trajectory data, the speed condition may be applied on the underlying network (e.g., the average passing speed of the nearest road crossing $\overline{crossingAvgSpeed}$ and the average passing speed of the map matched road segment $\overline{segmentAvgSpeed}$), instead of the $\overline{cellAvgSpeed}$. Algorithm 2.1 provides the pseudocode to determine Δ_{speed}. The sensitivity of the coefficients δ_1 and δ_2 (e.g., $\delta_1 = \delta_2 = \delta = 30\%$) can be analyzed through experiments to choose appropriate δ for computing stops.

Algorithm 2.1: getDynamicΔ_{speed} (gpsPoint, obj_{id}, δ)

---

input : gpsPoint $p = (x, y, t)$, moving object obj_{id}
output: dynamic speed threshold Δ_{speed}

1 get the average speed of this moving object obj_{id}: $\overline{objectAvgSpeed}$;
2 **if** *network-constrained trajectory* **then**
3 | get the average speed of the nearest road crossing to p: $\overline{crossingAvgSpeed}$;
4 | get the average speed of the map matched road segment of p: $\overline{segmentAvgSpeed}$;
5 | $positionAvgSpeed \leftarrow min\{\overline{crossingAvgSpeed}, \overline{segmentAvgSpeed}\}$
6 **else**
7 | get the average speed of the cell that (x,y) belongs to: $\overline{cellAvgSpeed}$;
8 | $positionAvgSpeed \leftarrow \overline{cellAvgSpeed}$
9 compute the dynamic speed threshold by Definition 2.4;
10 **return** Δ_{speed}

In some scenarios, GPS tracking data have instantaneous speed values (s) captured by the devices. They may be used for calculating Δ_{speed} and identifying the stops; otherwise, s is approximated by the average speed between the previous spatio-temporal point ($x_{i-1}, y_{i-1}, t_{i-1}$) and the next one ($x_{i+1}, y_{i+1}, t_{i+1}$), i.e., $s_i = \frac{\|\langle x_{i+1}, y_{i+1} \rangle - \langle x_{i-1}, y_{i-1} \rangle\|_2^2}{t_{i+1} - t_{i-1}}$. This is possible as GPS data is usually sampled frequently (e.g., few samples per min).

Algorithm 2.2 summarizes *velocity-based trajectory structure*. First, the instantaneous speed is computed if it is not available from GPS devices. Second, the dynamic Δ_{speed} (using Algorithm 2.1) is computed and the GPS point is annotated with "M" or "S" tag. Finally, stops and moves are computed by aggregating all consecutive points with the same tag, with a precondition on the minimal stop duration τ. This algorithm has linear complexity on the size of GPS feed, together with linear complexity on the size of road segments in the underlying network. It performs two data scans while tagging points and grouping consecutive points for the episodes. However, it is possible to combine the two scans together for better performance and shorten the computing time.

Density-based Trajectory Structure Identification

Using only velocity for identifying stops is not enough for some applications. For example when analyzing bird migrations, the foraging stops need to be found. Some birds, like water-birds, when they are looking for food, can fly at high speed, but inside a small area. Another example is in traffic applications, when someone is driving quickly around a block looking for a parking place. The velocity-based algorithm cannot detect these kinds of stops. Researchers have designed a *density-based* stop identification method, which considers not only the speed but also the maximum distance that the moving object has travelled during a given time duration. For this algorithm, density areas for extracting stop or move episodes need to be defined.

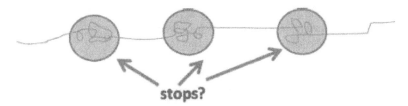

Figure 2.15: Density-based stop identification.

Definition 2.5 ($\mathcal{A}_{density}$ - **Density Area**)   Given a cleaned sequence of GPS points $\{\langle x_i, y_i, t_i \rangle\}$, a maximum distance σ, and a time duration τ, a density area $\mathcal{A}$ is a sub-sequence of the GPS points $\{\langle x_{i1}, y_{i1}, t_{i1} \rangle, \ldots, \langle x_{im}, y_{im}, t_{im} \rangle\}$ that satisfies two conditions.

1) For any two different points of the density area, if they are temporally distant by less than τ then they are spatially distant by less than σ, i.e., $\forall \langle x_{ia}, y_{ia}, t_{ia} \rangle, \langle x_{ib}, y_{ib}, t_{ib} \rangle \in \mathcal{A}$, $\|t_{ib} - t_{ia}\| \leq \tau$ $\Rightarrow \|\langle x_{ia}, y_{ia} \rangle - \langle x_{ib}, y_{ib} \rangle\| \leq \sigma$.

2) For the last *(first)* point of the GPS sequence that is just before *(after)* the density area, say $\langle x_b, y_b, t_b \rangle$ ($\langle x_a, y_a, t_a \rangle$), there exists a point inside the density area, which is temporally distant by less than τ and spatially distant by more than σ, i.e., $\exists \langle x', y', t' \rangle \in \mathcal{A}$ $\|t' - t_b\| \leq \tau$ and $\|\langle x', y' \rangle - \langle x_b, y_b \rangle\| > \sigma$ *($\|t_a - t'\| \leq \tau$ and $\|\langle x_a, y_a \rangle - \langle x', y' \rangle\| > \sigma$).*

Algorithm 2.2: Velocity-based trajectory structure

Input: a raw trajectory $\mathcal{T}_{raw} = \{p_1, p_2, \cdots, p_n\}$
Output: a structured trajectory $\mathcal{T}_{str} = \{e_1, e_2, \ldots, e_m\}$ where e_i is a tagged trajectory episode (stop $\mathcal{S}$ or move $\mathcal{M}$)

1 **begin**
2   /* initialize: calculate GPS instant speed if needed */
3   ArrayList$\langle x, y, t, tag \rangle$ $gpsList \leftarrow$ getGPSList($\mathcal{T}_{spa}$);
4   **if** *no instant speed from GPS device* **then**
5   compute GPS instant speed s_i for all $p_i = (x, y, t) \in gpsList$;
6   /* episode annotation: tag each GPS point with 'S' or 'M' */
7   **forall the** $p_i = (x, y, t) \in gpsList$ **do**
8   // *get dynamic* $\Delta_{speed}^{(i)}$ *by Algorithm 2.1*
9        $\Delta_{speed}^{(i)} \leftarrow$ getDynamicΔ_{speed} (p, obj_{id}, δ);
10   // *tag GPS point as a stop point 'S' or a move point 'M'*
11   **if** *instant speed* $s_i < \Delta_{speed}^{(i)}$ **then**
12   tag current point $p_i(x, y, t)$ as a stop point 'S';
13   **else**
14   tag current point $p_i(x, y, t)$ as a move point 'M';
15   /* compute episodes: grouping consecutive same tags*/
16   **forall the** *consecutive points with the same tag 'S'* **do**
17   // *compute stop episode*
18   get the total time duration $t_{interval}$ of these points;
19   **if** $t_{interval} > \tau$ *the minimal possible stop time* **then**
20          $stop \leftarrow (time_{from}, time_{to}, center, boundingRectangle)$;
21          $\mathcal{T}_{str}.(stop, \text{'S'})$; // *add the stop episode*
22   **else**
23   change the 'S' tag to 'M' for all these points; // *as "congestion"*
24   **forall the** *consecutive points with the same tag 'M'* **do**
25   // *compute move episode*
26       $move \leftarrow (stop_{from}, stop_{to}, duration)$ // *create a move episode*
27       $\mathcal{T}_{str}.(move, \text{'M'})$; // *add the move episode*
28   **return** the structured trajectory $\mathcal{T}_{str}$;

A variant of the well-known DBSCAN algorithm [50] may be employed for computing density-based stops. DBSCAN (Density-Based Spatial Clustering of Applications with Noise) is a well-cited spatial clustering algorithm proposed by Ester et al. in 1996, which looks for core points in order to start a cluster and then expand the clusters by adding density-reachable points. For extending DBSCAN to discover stops in trajectory data, we design TrajDBSCAN following the DBSCAN principles with certain modifications for trajectories. Since we want to find stops in a single trajectory with respect to both space and time factors, the main concern is that stop must be a continuous sub-sequence of a trajectory; hence, a stop should contain only time consecutive points. We also need to overcome the problems like GPS point absence because of signal loss or low sampling rate. The method looks for core points and then expands them by aggregating other points in the neighborhood. The main distinguishing design principles are the following.

- The *neighborhood* of a point p is determined not only by the spatial distance ε_{space}, but also the temporal information, i.e., it only considers the temporal linear neighborhood.

- In contrast to using the minimal number of points *minPts* for fostering clusters in DB-SACN, TrajDBSCAN determines whether p is a core point or not by the minimum stop duration $minTime$, not the *minPts*.

Algorithm 2.3 describes the detailed procedure of such density-based stop discovery method. First, we initialize an empty set of stop episodes as a structured trajectory needs to be computed (line 1); then the method iterates through the trajectory and processes the data points that have not yet been processed (from line 2). In line 5, the $neighbor_\varepsilon$ of the point is computed and checked with the ε_{time} constraint to examine if the point is a core-point (line 6). If a core-point is found, a new cluster (stop) is created (line 7). Afterward, the method aggregate other points in the neighborhood to expand the cluster (line 8) by the *expandStop* function, which recursively checks the points in the cluster to find the possible neighborhood cluster to expand the stop. Finally, the stop cluster is added to the output stop set (line 10).

Both velocity-based and density-based trajectory structure methods annotate each GPS point $\langle x, y, t \rangle$ with "M" or "S". Stops and moves are then computed based on contiguous "M"/"S" tags, together with the begin/end tags ("B"/"E") resulting from the trajectory segmentation layer. Thus, a continuous sequence of $\langle x, y, t \rangle$ points having all "M" tags is integrated into a single *move*, while, a continuous sequence of $\langle x, y, t \rangle$ points, all with "S" tags, is integrated into a single *stop*. The first and last $\langle x, y, t \rangle$ point of each trajectory are, respectively, computed as its *begin* and *end*.

2.4   SEMANTIC TRAJECTORY ANNOTATION

The trajectory computation layers developed different levels of data abstraction, reconstructed trajectories as a sequence of highly correlated episodes, resulting in structured trajectories $\mathcal{T}_{str}$. An important next step to better understand these episodes is to uncover the semantic meaning of each episode. For example, is this *stop* episode implying the user is at home? Is this *move* episode implying the user is using public transportation? We call a *Semantic Trajectory* as a trajectory that is enriched with information at this level. This has been defined earlier in the previous section about trajectory modeling. Figure 2.16 shows an example of a semantic trajectory. third-party geographic information sources like landuse distribution, road network from Google Map or Openstreetmap are needed for creating such semantic abstractions. In this section, we describe algorithms to compute such a model from GPS trajectories. Logically, this layer is on top of the Trajectory Structure layer. The algorithms can operate on top of the *stop* and *move* episodes.

Algorithm 2.3: Density-based stop discovery—TrajDBSCAN

---

    input : $\mathcal{Q} = \{Q_1, \cdots, Q_n\}$ //trajectory
              ε_{time} //the stop minimum time
              ε_{space} //the neighborhood distance for fostering a stop
   output: $\mathcal{T}_{str}$ structured trajectory as a set of stops w.r.t ε_{time} and ε_{space}

1 $\mathcal{T}_{str} = \varnothing$
2 **foreach** *point Q_i in $\mathcal{Q}$* **do**
3   **if** *Q_i is unprocessed* **then**
4   mark Q_i as processed;
5        $\mathcal{N} = neighbor_\varepsilon(Q_i, \varepsilon_{space})$;
6   **if** *duration($\mathcal{N}$) $> \varepsilon_{time}$* **then**
7   S = new *stop*; // as next stop discovered
8           $S = expandStop(Q_i, \mathcal{N}, S, \varepsilon_{space}, \varepsilon_{time})$;
9           $\mathcal{T}_{str} = \mathcal{T}_{str} \cup S$

10 **return** $\mathcal{T}_{str}$;

11 **function** *expandStop* $(Q_i, \mathcal{N}, S, \varepsilon_{space}, \varepsilon_{time})$
12 $S = \mathcal{N}$;
13 initialize the stop as the first $\mathcal{N}$ **foreach** *point Q_j in N* **do**
14   **if** *Q_j is unprocessed* **then**
15   mark Q_j as processed;
16        $\mathcal{N}' = neighbor_\varepsilon(Q_j, \varepsilon_{space})$;
17   **if** *duration(N') $> \varepsilon_{time}$* **then**
18           $S = S \cup \mathcal{N}'$

19   **return** S;

Our objective now is to provide a uniform and generic annotation framework for enriching structured trajectories with multiple geographic artifacts in order to transform them to semantic trajectories. We have a few challenges to achieve this objective. First, the framework should be able to cover the requirements of a wide range of applications. Second, multiple heterogeneous information sources need to be integrated together. Third, the GPS trajectories themselves and the data sources might have varying data quality. The framework should accommodate that and degrade gracefully. As a first move towards a semantic representation of trajectories, we introduce *Semantic Places (P)* as the semantic counterpart of the spatio-temporal positions. They denote geographic objects defined in or inferred from third-party information sources that contain data about the geographic objects of interest to the application at hand.

Definition 2.6   Semantic Places ($\mathcal{P}$) - A set of meaningful geographic objects used for annotating trajectory data. Each place *sp* has an extent and additional attributes containing useful metadata $(a_1, a_2, \cdots, a_n)$ for describing the place. The set $\mathcal{P}$ is partitioned into three subsets that

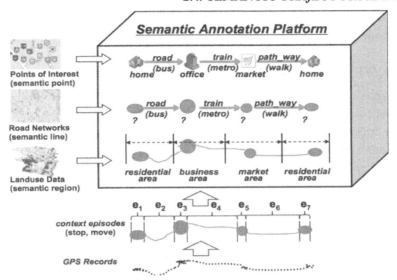

Figure 2.16: Trajectory annotation platform.

are defined according to the geometric shape of their extent, i.e., $\mathcal{P} = \mathcal{P}_{region} \bigcup \mathcal{P}_{line} \bigcup \mathcal{P}_{point}$, where:

- $\mathcal{P}_{region} = \{r_1, r_2, \quad , r_{n_1}\}$ is a set of places whose extent is a region;

- $\mathcal{P}_{line} = \{l_1, l_2, \quad , l_{n_2}\}$ is a set of places whose extent is a line; and

- $\mathcal{P}_{point} = \{p_1, p_2, \quad , p_{n_3}\}$ is a set of places whose extent is a point.

Region (or area), line, and point are standard spatial data types routinely used in GIS. Their formal definitions can be found in, e.g., [70]. For the rest of the chapter, we also denote $\mathcal{P}_{region}$; $\mathcal{P}_{line}$; $\mathcal{P}_{point}$ as Regions of Interest (ROI), Lines of Interest (LOI), and Points of Interest (POI).

Here, a layered approach has been adopted to construct the framework. A layered approach can be carefully designed to support efficient semantic annotation, considering the challenges. We first annotate the trajectory episodes with ROIs, i.e., regions on which the trajectory has passed. To do this, *spatial join* algorithms are applied for integrating geographic data with trajectory data. This gives a coarse-grained view of the semantic movement. Thereafter, a *semantic line annotation* algorithm is designed that annotates *move* episodes, computing $\mathcal{T}_{sem}^{(line)}$ (a sequence of semantic moves) using LOIs (e.g., road network). For $\mathcal{P}_{point}$, a *hidden Markov model* (*HMM*) based algorithm is designed for annotating *stop* episodes, computing $\mathcal{T}_{sem}^{(point)}$ with POIs (i.e., home, office, shopping mall, restaurant, etc.). Wherever appropriate, the algorithms are redesigned (particularly for line and point annotation) considering the challenge that algorithms should exhibit good performance over a wide range of trajectories with varying data quality. Next, we provide an overview each of these annotation methodologies.

2.4.1   ANNOTATION WITH SEMANTIC REGIONS

This layer enables annotation of trajectories with meaningful geographic regions. It does so by extracting data from several third-party data sources containing information about the regions on which the trajectory has passed. As discussed before, regions are technically referred to as semantic regions having a spatial extent ($\mathcal{P}_{region}$), i.e., an insignificant coverage area, that makes it different from a spatial point, for e.g., a shop. The algorithms used to perform such data extraction are called *spatial join* algorithms.

Spatial join algorithms compute a topological correlation between the trajectory data $\mathcal{Q}$ and semantic regions $\mathcal{P}_{region}$ (i.e., $\mathcal{Q} \bowtie_\theta \mathcal{P}_{region}$). θ is referred to as the spatial predicate, i.e., the type of join syntax employed. This depends on the type of data we are attempting to join the trajectory with. Generally, spatial join can compute the correlations about *directional*, *distance*, and *topological* spatial relations such as *intersection* and *union* [19]. Taking the *stop* episodes for example, spatial subsumption (ObjectA is *inside* ObjectB) are found to be the most used predicate. For the spatial extent, either the spatial *bounding rectangle* of the episode (for move or stop) or its *center* (for stop) to perform spatial join is used. After finding the appropriate regions (r_i), each episode in the trajectory is annotated with these regions and its associated metadata.

These semantic regions can be free form regions like an university campus, a recreation facility with a swimming pool or grids consisting of highly regular cells. Different data sources have different types of regions. For example, Openstreetmap[16] can have free-form regions and Swisstopo[17] landuse details are highly regular in the form of cells and zones.

Figure 2.17 shows one person's trajectory on Sunday, annotated with semantic places of various kinds taken from Swisstopo (residential area, recreational area) and Openstreetmap (EPFL campus). By using an application database (e.g., EPFL's employee database) annotations for this personal trajectory can be expressed as: *home* → *EPFL campus (staying 4 hours)* → *a swimming pool (staying 1 hour)* → *home*. As an example of landuse-related information sources, Figure 2.18 illustrates landuse classification categories and subcategories that Swisstopo uses to annotate 1,936,439 cells (100 m × 100 m) covering Switzerland. Figure 2.19 is an example of an annotated trajectory with such landuse cells.

Algorithm 2.4 shows the pseudocode of the annotation algorithm with regions, which directly annotates GPS records with regions. Note that, depending on requirements, the spatial join can be computed only for selected episodes. R*-tree index is applied on semantic regions $\mathcal{P}_{region}$ [10] to improve efficiency of the algorithm. The complexity of the annotation algorithm with region is $O(n * log(m))$, where n is the number of GPS records (or stop episodes) while m is the size of $\mathcal{P}_{region}$. For well-divided landuse data, the complexity can be even less, i.e., $O(n)$.

---

[16]http://www.openstreetmap.org
[17]http://www.swisstopo.admin.ch/

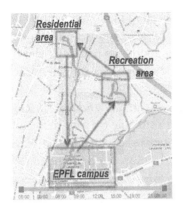
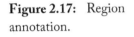

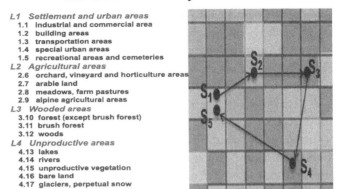

Figure 2.17: Region annotation.

Figure 2.18: Landuse ontology.

Figure 2.19: Synthetic data or landuse.

L1 *Settlement and urban areas*
  1.1 industrial and commercial area
  1.2 building areas
  1.3 transportation areas
  1.4 special urban areas
  1.5 recreational areas and cemeteries
L2 *Agricultural areas*
  2.6 orchard, vineyard and horticulture areas
  2.7 arable land
  2.8 meadows, farm pastures
  2.9 alpine agricultural areas
L3 *Wooded areas*
  3.10 forest (except brush forest)
  3.11 brush forest
  3.12 woods
L4 *Unproductive areas*
  4.13 lakes
  4.14 rivers
  4.15 unproductive vegetation
  4.16 bare land
  4.17 glaciers, perpetual snow

Algorithm 2.4: Trajectory annotation with ROIs

Input: (1) a raw trajectory Q with its sequence of GPS points $\{Q_1, \dots, Q_n\}$, (2) a set of semantic regions $\mathcal{P}_{region} = \{region_1, \dots, region_{n_1}\}$

Output: structured semantic trajectory $\mathcal{T}_{region}$

1 **begin**
2    $\mathcal{T}_{region} \leftarrow \emptyset$; //initialize the trajectory
3   /* compute *intersections* between Q and $\mathcal{P}_{region}$; */
4   do spatial joins $Q \bowtie_{intersect} \mathcal{P}_{region}$;
5   /* process each *intersection* and compute trajectory tuple */
6   **forall the** *intersected regions* **do**
7   group continuous GSP point $Q_i \in Q$ in the *intersection*;
8   approximate entering time t_{in} and leaving time t_{out};
9   create a trajectory *tuple* $\leftarrow (region_j, t_{in}, t_{out}, reg_{type})$;
10   **if** *current* reg_{type} *= previous* reg_{type} **then**
11   ;
12   merge the two tuples into a single tuple; **else** $\mathcal{T}_{region}$.add(*tuple*); //add the previous tuple to $\mathcal{T}_{region}$;
13   ;
14    $\mathcal{T}_{region}$.add(*tuple*); //add the last tuple to $\mathcal{T}_{region}$;
15   **return** trajectory $\mathcal{T}_{region}$

2.4.2 ANNOTATION WITH SEMANTIC LINES

This layer annotates trajectories with Lines Of Interest (LOI) and considers variations present in heterogeneous trajectories (e.g., vehicles run on road networks, human trajectories use a combination of transport networks and walk-ways, etc.). Given data sources of different forms of road networks, the purpose is to identify the *correct* road segments as well as infer *transportation modes* such as *walking, cycling, public transportation like metro*. Thus, the algorithms in this layer consist of two major modules: the first module focuses on designing good map matching algorithms to identify the correct road segments for the move episodes. The second one module focuses on inferring the transportation mode that the moving object used.

Map-matching algorithms usually design a distance metric (e.g., *perpendicular distance*) to map the GPS points to the nearest road segment [134]. Although suitable for well-defined highway networks, perpendicular distance is not suitable for dense networks, parallel road-ways and arbitrary crossings. This is because vertical projections of (x,y,t) points on corresponding road segments often do not fall on the segment. To overcome this, the *point-segment distance* is applied, defined as:

$$d(Q, A_i A_j) = \begin{cases} d(QQ') & \text{if } Q' \in A_i A_j \\ \min\{d(QA_i), d(QA_j)\} & \text{otherwise} \end{cases}, \qquad (2.3)$$

where Q' is the projection of the GPS point Q on the line determined by the two crossings A_i and A_j; $d(QQ')$ is the perpendicular distance between Q and that line; $d(QA)$ is the Euclidean distance between Q and the crossing A.

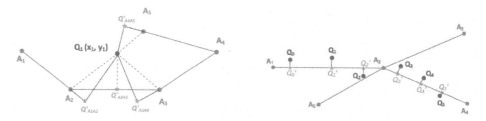

Figure 2.20: Point-segment distance.   **Figure 2.21:** Global map-matching.

As a subsequence of raw trajectory Q, a *move* episode also includes a list of spatio-temporal points. Choosing the candidate road segment for each single point independently sometimes results in incorrect mapping, specially for non-perpendicular path ways. Global map matching algorithms have shown better matching quality [18][134] as they consider the context of neighboring points to determine the road segment. Let us adopt this, along with the *point-segment distance*, in order to design two metrics (*localScore* and *globalScore*) to map *move* episodes to appropriate road segments for heterogeneous road structures.

Consider a *global view radius R* around candidate points, with a context window of size $2R$. Therefore, mapping results of point Q depend also on the effects of its neighboring points (N_1 points before and N_2 points after in radius R). For computational efficiency, only the *neighboring* segments are considered as candidate road segments *candidateSegs(Q)*. They can be efficiently accessed with R*-tree index [10]. The point-segment distance $d(Q, A_i A_j)$ is normalized as the *localScore* between point Q and road segment $A_i A_j$.

$$localScore(Q, A_i A_j) = \begin{cases} \frac{d_{min}(Q)}{d(Q, A_i A_j)} & A_i A_j \in candidateSegs(Q) \\ 0 & otherwise \end{cases} \quad (2.4)$$

where $d_{min}(Q)$ is the shortest distance from Q to all possible candidate road segments $A_i A_j$. Based on *localScore*, a global measurement—*globalScore*—is computed between Q and $A_i A_j$ considering the context window $2R$ containing N_1 points prior to Q and the forthcoming N_2 points:

$$globalScore(Q, A_i A_j) = \frac{\sum_{k=-N_1}^{N_2} w_k \cdot localScore(Q_k, A_i A_j)}{\sum_{k=-N_1}^{N_2} w_k} \quad (2.5)$$

$$w_k = \begin{cases} \exp(-\frac{d(Q_0 Q_k)^2}{2\sigma^2}) & d(Q_0 Q_k) < R \\ 0 & otherwise \end{cases}, \quad (2.6)$$

where Q_k is the kth neighboring point of Q (e.g., Q_0 is Q itself, Q_{-1} is the previous point whilst Q_{+1} is the next point); w_k is the corresponding weight determined by a Kernel smoothing function with the Kernel bandwidth σ.

After this first step of the global map matching, each episode is annotated in terms of a list of road segments, i.e., $ep = \{r_1, r_2, \ldots, r_l\}$. Thereafter, the annotation of transportation mode is conducted on each segment (or route). As a result, we obtain pairs of $\langle r_i, mode_i \rangle$ for each annotated episode falling on road segment r_i. These annotations are determined using simplistic rules applied to the characteristics of the "annotated" move episode. Characteristics of move episode could be statistical properties of the episode such as *average velocity*, *average acceleration*, *road type*, etc.

Algorithm 2.5 shows the detailed procedure of semantic line annotation: (1) select candidate road segments; (2) calculate the point-segment distance; (3) normalize the distance as *localScore*; (4) compute the weight and calculate *globalScore*; (5) determine the map matching segment for each point based on *globalScore*; and (6) further infer the transport mode based on the features of the segment and the road type information.

Since each GPS point considers only the neighboring road segments as a set of candidate segments, the candidate set size is significantly smaller than the total size of road networks in real-life datasets. This makes the algorithm, besides having better matching quality, also efficient, with linear complexity in the order of the GPS points $O(n)$. The global map matching parameters (e.g., radius R and kernel width σ) are tuned for different experiments.

As an example, Figure 2.22 shows a user's typical *home-office* trajectory. It shows that the user walked a few blocks from home, then took the public transportation—metro—and finally walked from the metro stop to his office: sub-figure (a) shows the original GPS trace; (b) displays the initial map-matched road segments for these GPS points; (c) further infers the corresponding different transportation modes such as *metro* or *walk*; and finally (d) summarizes the mobility trace in terms of sequences of roads that are stored in the semantic trajectory store.

Algorithm 2.5: Trajectory annotation with LOIs

Input: (1) a move episode of raw trajectory Q of GPS points $\{Q_i(x_i, y_i, t_i)\}$
　　　　(2) a set of road segments $P_{line} = \{r_1, r_2, \cdots, r_m\}$
Output: semantic trajectory $\mathcal{T}_{line}$

1 **begin**
2 　　$preSeg \leftarrow \emptyset$, $\mathcal{T}_{line} \leftarrow \emptyset$; //initialize the trajectory
3 　　**forall the** $Q_i = (x, y, t) \in Q$ **do**
4 　　　　/* **select candidate roads for** Q_i **(R*-tree)***/
5 　　　　$candidateSegs(Q_i) \leftarrow \{r_1^{(i)}, \cdots, r_n^{(i)}\}$; // select only neighboring road segments
6 　　　　/* **calculate dist., normalize it as localScore** */
7 　　　　compute the distance between point Q_i and $\forall r_j^{(i)} \in candidateSegs(Q_i)$;
8 　　　　choose the closest segment $min\{d(Q_i, r_j^{(i)})\}$ (Equ. 2.3);
9 　　　　normalize distance as $localScore(Q_i, r_j^{(i)})$ $\forall r_j^{(i)} \in candidateSegs(Q_i)$ by Formula 2.4;
10 　　　　/* **calculate globalScore: (point, segment)** */
11 　　　　choose global points $(Q_{-N_1}, \cdots, Q_{+N_2})$ in radius R;
12 　　　　compute their Kernel smoothing weights by Formula 2.6;
13 　　　　compute the $globalScore(Q_i, r_j^{(i)})$ for $\forall r_j^{(i)} \in candidateSegs(Q_i)$ by Formula 2.5;
14 　　　　/* **compute** Q' **with road position (if needed)** */
15 　　　　rank the computed $globalScore(Q_i, r)$
16 　　　　choose the highest score to match $segmentId$ for Q_i;
17 　　　　compute the corrected position (x', y') if needed ;
18 　　　　/* **add road segment as a trajectory tuple** */
19 　　　　**if** $preSeg \neq null$ and $preSeg \neq segmentId$ **then**
20 　　　　　　/* **infer transportation mode** */
21 　　　　　　get $transportMode$ by velocity distribution, road information, etc.
22 　　　　　　/* **add the semantic episode** */
23 　　　　　　$(segmentId, time_{in}, time_{out}, mode) \rightarrow \mathcal{T}_{line}$;
24 　　　　　　$preSeg \leftarrow segmentId$;
25 　　**return** structured semantic trajectory $\mathcal{T}_{line}$

2.4.3   ANNOTATION WITH SEMANTIC POINTS

This layer annotates the *stop* episodes of a trajectory with information about plausible *points of interest* (POIs). Examples of POI are *Gino restaurant*, *Armani shop Via Manzoni*, etc. For scarcely populated areas, it is trivial to identify the POI that is the goal of a stop (e.g., the goal of a stop at a highway petrol pump is the petrol pump itself). However, densely populated urban areas (e.g., downtown or old town areas) may have many candidate POIs for each stop. Further, low GPS sampling rate due to battery outage and signal losses makes the problem intricate.

For instance, some data sets have overwhelmingly large number of POIs with largely varying density. This large number makes it probabilistically intractable to infer the exact POI of the stop from imprecise location records. Instead, it is more feasible to try to extract some seman-

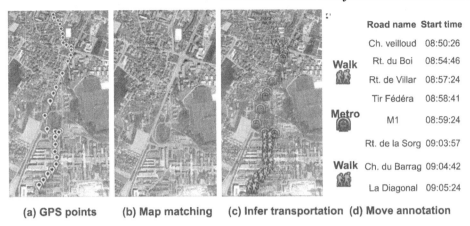

	Road name	Start time
Walk	Ch. veilloud	08:50:26
	Rt. du Boi	08:54:46
	Rt. de Villar	08:57:24
	Tir Fédéra	08:58:41
Metro	M1	08:59:24
	Rt. de la Sorg	09:03:57
Walk	Ch. du Barrag	09:04:42
	La Diagonal	09:05:24

(a) GPS points   (b) Map matching   (c) Infer transportation   (d) Move annotation

Figure 2.22: Move annotation: a home-office move via taking metro and walking.

tic characteristics of the stop using aggregate information about the surroundings, that could be important for heterogeneous applications. For example, instead of inferring the exact restaurant a trajectory traverses, the inference can be more reliable with just the probable semantic activity associated with the stop (e.g., eating, shopping). We describe such a method next. We use an example dataset from Milan for this purpose [62]. In this dataset, POIs are organized into a hierarchy according to their category for the local administration. The top level of the hierarchy contains five generic categories: *services, food, home_item, personal_item,* and *other*. Our objective would be to infer these *categories* for stop episodes recorded within Milan.

The technique is based on *Hidden Markov Models* (*HMM*). It has been designed for the semantic annotation of *stops* with POI category. An unique novelty of the approach is that it works for densely populated areas with many possible POI candidates for annotation, thus catering to heterogeneous people and vehicle trajectories. Algorithms that focus on identification of the exact POI, typically working better with datasets having sparser POIs can be found in [5, 162].

HMM is a classical statistical signal model in which the system being modeled is assumed to be a Markov process with unobserved state [135]. We consider the *temporal sequence* of GPS stops: $\mathcal{S} = (S_1, S_2, \cdots, S_n)$ as the observed values.

Figure 2.23 expresses the resultant HMM problem. The initial input is the raw trajectory $\mathcal{Q}$, i.e., the sequence of (x,y,t) points; a *sequence of stops* is computed and forms the real observation (O); and the POI *instances* are the superficial hidden states, while the POI *categories* are the real hidden states that we are interested in. The goal is to identify the real hidden states and use them to annotate the stops.

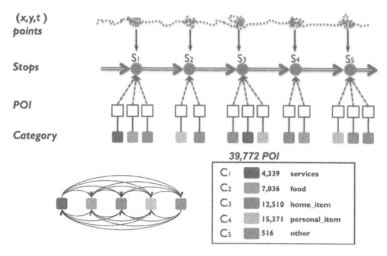

Figure 2.23:   HMM formalism for inferring POI category.

HMM-based Trajectory Modeling

Let there be m POI categories $C_1 \dots C_m$. Typically, a HMM λ has three major components, i.e., $\lambda = (\pi, \mathcal{A}, \mathcal{B})$, where π is the probability of the initial states, i.e., $Pr(C_i)$, $\mathcal{A}$ is the state transition probability matrix $([Pr(C_j|C_i)]_{m \times m})$, $\mathcal{B}$ is the observation probability for each state $Pr(o|C_i)$.

- **Initial Probabilities (π).** We approximate the probability of initial states π as the percentage of POI samples belonging to each category from the information source. Therefore, for Milan POI dataset,

$$\pi = \left\{ \frac{4339}{39772}, \frac{7036}{39772}, \frac{12510}{39772}, \frac{15371}{39772}, \frac{516}{39772} \right\}.$$

- **State Transition ($\mathcal{A}$).** State transition probability $Pr(C_j|C_i)$ in the formulation represents the possible sequences of stop categories, i.e., the probability to stop in a POI of category C_j given that the previous stop was in a POI of category C_i. Wherever available, category sequences (e.g., *food → items for people* or *food → other*) are obtained through other information sources (e.g., from *region* transitions). For trajectories having insufficient history, the state transition matrix is initialized following nomenclatures of the POI categories (e.g., associate high probability for meaningful state transitions and low probabilities for non-meaningful state transitions).

$$\mathcal{A} = \begin{pmatrix} 0.8 & 0.05 & 0.05 & 0.05 & 0.05 \\ 0.05 & 0.8 & 0.05 & 0.05 & 0.05 \\ 0.05 & 0.05 & 0.8 & 0.05 & 0.05 \\ 0.05 & 0.05 & 0.05 & 0.8 & 0.05 \\ 0.15 & 0.15 & 0.15 & 0.15 & 0.4 \end{pmatrix}$$

- **Observation Probabilities ($\mathcal{B}$).** $Pr(o|C_i)$ intuitively represents the probability of seeing a *stop o* (as the observation) in $\mathcal{T}$ *caused by* user's interest in places belonging to category C_i. $Pr(o|C_i)$ can be approximated by using the center of the *stop* $Pr(center_{xy}|C_i)$ or the bounding rectangle $Pr(boundRectangle|C_i)$.

Computing $\mathcal{B}$ for areas having high POI density is not easy. The solution is based on the intuition that the influence of a POI category on a *stop* is proportional to the number of POI instances of that category in the stop area. The influence of a POI is modeled as a two-dimensional Gaussian distribution—the mean is the POI's physical position (x, y) and the variance is $[\sigma_c^2, 0; 0, \sigma_c^2]$, where σ_c is category specific. Figure 2.24 displays an example of 12 POIs' Gaussian distributions with the corresponding densities in Figure 2.25. By Bayesian rule, the lemma is deduced to determine $Pr(o|C_i)$ in $\mathcal{B}$.

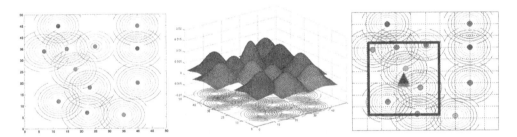

Figure 2.24: POI distribution.

Figure 2.25: POI densities.

Figure 2.26: POI discretization.

Lemma 1 *$Pr(o|C_i)$ is proportional to the sum of the probability of each POI that belongs to this category C_i, namely $Pr(o|C_i) \propto \Sigma_j Pr(o|poi_j^{(C_i)})$.*

Proof.

$$\begin{aligned} Pr(o|C_i) &= \frac{Pr(o, C_i)}{Pr(C_i)} = \frac{\Sigma_j Pr(o, poi_j^{(C_i)})}{\Sigma_j Pr(poi_j^{(C_i)})} = \frac{\Sigma_j Pr(o|poi_j^{(C_i)}) Pr(poi_j^{(C_i)})}{\Sigma_j Pr(poi_j^{(C_i)})} \\ &\propto \Sigma_j Pr(o|poi_j^{(C_i)}) Pr(poi_j^{(C_i)}) \propto \Sigma_j Pr(o|poi_j^{(C_i)}) \end{aligned}$$

$\square$

Discretization and *neighboring* techniques are employed to improve the efficiency of computing $Pr(o|C_i)$. Using *discretization*, the area of POIs are divided into grids (jk) and the discrete probability values of $Pr(grid_{jk}|C_i)$ are precomputed, as the approximation of $Pr(center_{xy}|C_i)$. Further, for each $grid_{jk}$, only *neighboring* POIs in that box (black rectangle in Figure 2.26)are considered, instead of all the POIs in the area.

Inferring Hidden States—Semantic Stops

Using the above defined complete form HMM $\lambda = (\pi, \mathcal{A}, \mathcal{B})$, their hidden states (the purpose behind the stops) $HS = \{pc_1, pc_2, \cdots, pc_n\}$ may be inferred from the stop sequence $OV = \{stop_1, stop_2, \cdots, stop_n\}$ available through the stop/move computation phase; where pc_t is the POI category $pc_t \in \{C_1, \cdots, C_m\}$. This problem can be formalized as maximizing the likelihood $\mathcal{L}(HS|OV, \lambda)$. This problem is redefined as a *dynamic programming* problem, defining $\delta_t(i)$ as the highest probability of the tth *stop* caused due to POI category C_i (Formula 2.7). Formula 2.8 gives the corresponding induced form of highest probability at the $(t + 1)^{th}$ stop for category C_j, considering the state transition probabilities. The previous state C_i that gives the highest probability is recorded to current state C_j by $\psi_{t+1}(j)$ (Formula 2.9):

$$\delta_t(i) = \max_i Pr(pc_1, \cdots, pc_t = C_i, o_1, \cdots, o_t|\lambda) \tag{2.7}$$

$$\delta_{t+1}(j) = \max_i\{\delta_t(i)A_{ij}\} \times B_j(o_{t+1}) \tag{2.8}$$

$$\psi_{t+1}(j) = \operatorname*{argmax}_i \delta_t(i)A_{ij}. \tag{2.9}$$

Finally, the Viterbi algorithm [55] is employed to solve this dynamic programming problem for inferring the hidden state (stop category) sequence First $\delta_t(i)$ is recursively computed, and then the final stop state is deduced from the highest probability in the last stop. This is followed by a recursive backtracking to the previous stop state by $pc_{t-1}^* = \psi_t(pc_t^*)$. The details of the algorithm for inferring hidden stop category sequence is in Algorithm 2.6. The output of this layer is a sequence of semantic episodes describing the stops.

Finally, the results from the three annotation algorithms (regions, lines, points) are typically combined to produce the final semantic trajectory, which is exposed to applications. The semantic trajectory as a result now contains rich information about the semantic nature of the movement. For example, it contains data on the different regions that it has traversed; additional annotations on the move and stop episodes are available on a best effort basis. This data abstraction now becomes useful for heterogeneous applications.

2.5   SUMMARY AND OUTLOOK

In this chapter, we discussed the conceptual architecture and relevant techniques of computing semantic trajectories from positioning data like GPS feeds. Two semantic trajectory models are presented: *the trajectory ontological framework* and *the hybrid semantic trajectory model*. The former

---

Algorithm 2.6: Trajectory annotation with POIs

Input: (1) an observation sequence of stops
$O = \{Stop_1, Stop_2, \cdots, Stop_n\}$; (2) points of interest
$POIs = \{\langle p_1, q_1 \rangle, \cdots, \langle p_k, q_k \rangle\}$ where $q_i \in \{C_1, \cdots, C_5\}$
Output: a hidden state sequence about stop behaviors (in terms of POI categories), i.e.,
$S = \{q_1, q_2, \cdots, q_n\}, q_i \in \{C_1, \cdots, C_5\}$

1 **begin**
2   /* **learn the model from POIs** */
3     $\lambda = (\pi, \mathcal{A}, \mathcal{B})$
4   /* **initialization** */
5   **forall the** *POI category* C_i **do**
6        $\delta_1(i) = \pi_i B_i(o_1), 1 \leq i \leq N; \psi_1(i) = 0$
7   /* **recursion** */
8   **forall the** *t: 2 to n* **do**
9   **forall the** *categories* C_j **do**
10          $\delta_t(j) = \max_i[\delta_{t-1}(i)A_{ij}] \times B_j(o_t)$
11          $\psi_t(j) = \arg\max_i[\delta_{t-1}(i)A_{ij}]$
12   /* **termination** */
13    $P^* = \max_i[\delta_T(i)]; q_n^* = \arg\max_i[\delta_T(i)]$
14   /* **state sequence backtracking** */
15   **forall the** *t: n to 2* **do**
16       $q_{t-1}^* = \psi_t(q_t^*)$
17   /* **get the semantic trajectory with POI tags** */
18    $S = \{\langle stop_1, q_1 \rangle, \cdots, \langle stop_n, q_n \rangle\}$
19   summarize $\mathcal{T}_{point}$ from extracted POI sequence ($\langle stop, t_{in}, t_{out}, tagList \rangle$).
20   return structured semantic trajectory $\mathcal{T}_{point}$

---

one has the advantage of rich trajectory reasoning and queries, while the latter one is more dedicated to supporting the computational infrastructure for generating semantic trajectories from the raw sensor records. The applicability and choice of the two frameworks depends on the requirement, i.e., whether for high reasoning power or high semantics extraction capability.

We presented a comprehensive procedure to support progressive construction of different levels of trajectories, and the enrichment of trajectory semantics from multiple third-party semantic sources. It includes the computational steps (like data preprocessing such as cleaning and compression, trajectory identification, and trajectory structure with segmentation algorithms) and the annotation steps (e.g., semantic enrichment with various geographic and application knowledge like road network, point of interests).

We believe future works around building semantic trajectories from positioning data has a few interesting challenges and directions.

1) *Online computation of semantic trajectories*: Most of existing trajectory computing techniques work in an "offline" manner. To elaborate further, they work only when complete data of the trajectory is already available. It would be appealing to design online versions of these algorithms where, the annotation is performed while the data is streaming into the platform. With the increasing computational and storage capabilities in commodity devices (e.g., smartphones, tablets), we believe online algorithms and computational stacks to abstract semantic data from raw sensor data would take prominence and find wide-spread applicability.

2) *Indoor Spaces*: GPS is a power-hungry sensor and along with common activities on a commodity device, it can easily drain the power, thereby causing inconvenience to the users. Moreover, it does not work well in indoor spaces. Indoor spaces, on the other hand, have other techniques to help in localization. For example, WiFi signals can be used to locate and track an user. However, the quality is not that high and there are companies producing specialized chips for indoor positioning for this purpose. Mining and analyzing indoor trajectory feeds can be an interesting area of exploration, specially given its density attributes, compared to outdoor positioning.

3) *Validation and Automatically Labeling:* Although data sources for performing semantic annotations is on the rise, it is still a challenge to find an appropriately rich semantic source to correctly annotate a trajectory. Data qualities keep varying from source to source. Second, lack of ground-truth data is another issue. This makes validation of the annotations a big challenge. If an algorithm has labeled a stop as "shopping," how do we validate whether the prediction is correct? Most validation techniques today are indirect and empirical in nature. Since collection of such data is non-trivial, there is a need to develop automatically labeling techniques using un-supervised or semi-supervised algorithms with partially annotated ground truth data, e.g., data from FourSquare checkins and Twitter.

CHAPTER 3

Semantic Activities from Motion Sensors

The previous chapter focused on analyzing data obtained from GPS and provided a set of techniques for computing semantic trajectories from those GPS feeds. GPS is quite effective for tracking outdoor movements. As a consequence, trajectory mining primarily investigates how to extract meaningful information by combining the trace data with underlying geographic information systems, analyzing patterns that indicate various kinds of semantic mobility annotations like "stops," "moves," etc.

GPS, however, is not suitable for detecting micro-movement of users. Examples of such states are *walk, sit, stand, jump, run,* etc., although some of these activities result in a displacement of the user. This is because inherently GPS trajectories only catch the displacement-related parameters. It is not capable of detecting limb movements. Although one can argue that we can use GPS to detect whether a user is *walking on the street,* the accuracy of detection can be significantly strengthened if we knew that the person was *taking steps.* Equipped only with a GPS sensor, one cannot distinguish for example between a cyclist who is cycling at a very slow pace vs. a user who is walking. The purpose of motion sensors installed on different body parts of subjects is to precisely catch different ambulatory movements of users. This can then be used to infer the nature of micro-level movements the user is performing.

A typical city-bred office-going person spends a significant amount of time in indoor environments. Several micro-movements are manifested in indoor environments. For example, users are *sitting* at their desk in an office, *running* in a gym, *loitering* in a mall or a grocery store. In addition, users exhibit these micro-movements in a diverse number of ways. For example, one user can have a steady and straight strides while walking, while another may have staccato movements, for example, when he is browsing products in a store. The associated baggage a user is carrying (e.g., is he pushing a trolly?) also impacts the manifestation of the same micro-movement. Motion sensors are well suited to capture such personalized variations of micro-movements.

Along with micro-movements, users also carry out longer-duration activities. Examples of such activities are *cook, eat food, work-out in gym,* etc. Typically, these activities are called "complex activities." Researchers over the years have used different terminologies to describe different types of activities. Movements like *walk, sit, stand,* etc., usually have a repetitive sequence and are referred to as micro-movements or locomotive activities or micro-activities. Activity recognition researchers who focus primarily on these types of movements also refer to them as an "activity."

A complex activity, however, usually refers to activities that are long-running, and contains an uncertain mixture of micro-movements. In this chapter, we continue using the term "semantic activity" to refer to both the class of activities. This chapter will summarize techniques invented by researchers to learn the structures of these activities and to detect them.

Semantic activities mined from motion sensors hold significant promise for several business applications. The first application area is in healthcare. Motion sensors installed in different body parts can help users record their life-logs of exercise routines, amount of time spent doing physical activities like walking, cycling, etc. This information may be handy to provide recommendations to users of diet routines to follow. Elderly care agencies actively want to monitor their inmates to immediately detect falls, and also monitor different activity levels of users. This helps them monitor and provide feedback and exercise charters to their clients. Apart from these, indoor activities, in general, promise to have significant business applications in other domains such as retail, business gatherings, etc. [7, 46, 97, 153]. There is an increasing interest among researchers to understand how a crowd can be monitored on a large scale to provide solutions to businesses wanting to mine crowd behaviors (e.g., How do people shop?; Is there an emergency situation?). Motion sensors fall among one of the core sensors to be monitored in many such situations.

3.1   THE MOTION SENSOR

Monitoring of motion-based activities are achieved with the *accelerometer* (abbreviated as ACC), *gyroscope,* and *compass* [94]. Our focus remains on presenting the key techniques for mining semantics from motion sensors. While we would focus on accelerometers, the generic techniques for semantics mining are sensor independent.

3.1.1   SENSOR FUNCTIONALITY

The term "accelerometer" is very commonly understood today as a sensor on the phone. Most accelerometers are built using Micro-Electro-Mechanical System (MEMS) motion sensors.[1] They measure the *proper acceleration*[2] experienced by the object in which the accelerometer is embedded. Historically, accelerometers have been around for over a century. However, only recently, smartphone manufacturers have adopted this sensor and now it has reached almost every home, and every pocket. The sensor records the proper acceleration along three dimensions (X, Y, and Z), as shown in the low-level of Figure 3.1. The three mutually perpendicular axes are usually aligned with the phone. Hence, when the phone tilts, the reference axes tilts as well, with respect to an absolute coordinate system. If the phone is placed on a desk, horizontal to the ground, one axis (e.g., Z axis) points towards the center of the earth. Typically, the reference axes aligned with the phone's orientation is called the *local* coordinate system and the absolute system is called the *global* coordinate system [170].

---

[1]https://www.mems-exchange.org/MEMS/what-is.html
[2]http://en.wikipedia.org/wiki/Proper_acceleration

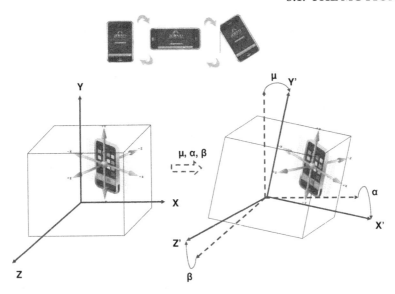

Figure 3.1: The local and global coordinate system. [X,Y,Z] is the global coordinate system and [X',Y',Z'] is the local coordinate system. θ, ϕ, ρ are the angles of shift.

The three axes record the acceleration experienced by the object along each dimension. When the phone is on a table, one of the axes records the earth acceleration (i.e., $g = 9.81m/s^2$) as a negative of the gravity vector ($\vec{g}$). The other two dimensions (say X and Y) have no acceleration in this stationary condition, except some noise, introduced in the measurement due to sensitivity issues of the MEMS sensor. The range of an accelerometer typically varies between $\pm 2g$.

Apart from accelerometers, the gyroscope provides angular momentum, which is often helpful to determine orientation changes on a smartphone. The gyroscope is widely used by smartphone gaming applications. In this chapter, we focus in depth, on semantics we can learn from the accelerometer sensor. This is because it is the primary sensor that has been widely studied by researchers to observe motion-inducing activities.

3.1.2   WHAT CAN WE LEARN FROM MOTION?

Several activities can be learnt from motion pattern of users. Researchers have worked with primarily two classes of activity semantics: (1) simple activities (also known as micro activities) and (2) complex activities (macro activities). Simple activities [9, 139] refer to activities that usually have a uniform repeatable pattern across several instances of the activity. Examples are walk, sit, stand, jump, etc. If a user walks for a while, the steps repeat and if we split the time series into multiple segments, we observe a repetitive pattern. The segment sizes typically are of the order of

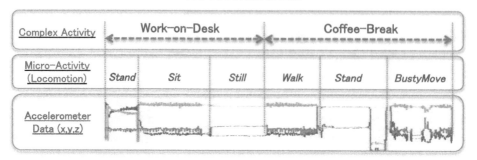

Figure 3.2: From accelerometer to semantic activities.

seconds. One can also understand that locomotive states would largely be falling under this class of activity semantics.

Complex activities [59, 66, 137] refer to semantic activities that typically span across a longer duration, is aperiodic with respect to patterns. Examples of such activities are "coffee-break" or "work-at-desk." These are also referred to as *activity routines* by some researchers. Semantic activities of this sort usually do not show repeatability of patterns across instances. However, there might be sufficient difference between two distinct complex activities (say "cook" and "dance"). Figure 3.2 provides an overview of the logical positioning of the two semantic concepts.

Research works typically select a set of activities and conduct their work on the selected set. The following are a few examples from different research works.

- [173] investigates the following activities: lying down, sitting, standing, household chores, walking, running, playing basketball, and dancing.

- [48] experiments with 5-min samples of the following activities: lying, sitting, standing, walking, running and cycling with an exercise bike.

- [98] conducted experiments with 1-min recordings per activity. The activities were [breakfast, lying, sitting, sleep, standing, still] (under activity class: idle/still); walking (under class: walk); running (under class: running); skiing (under class: skiing); bicycling (under class: cycling); [car, subway train, taxi, train](under class: vehicle); [cleaning, cooking, skating](under class: other).

- [28] focus on studying whether accelerometers can be used to differentiate between gait patterns of young and old adults, while climbing stairs.

Biomedical healthcare [87] is one important application area of activity recognition. As such, researchers in biomedicine have also investigated semantic activity recognition capacity of accelerometers, for example to detect abnormal activities like fall, stumbling, etc.

Figure 3.3 provides a exemplary dictionary of activities of daily living (referred to as ADL). Not all activities may be detected accurately using only the accelerometer. However, motion patterns often play an important role towards improving the confidence of learning algorithms to detect an activity.

3.1.3 DATA COLLECTION

Data collection refers to the methodology of collecting relevant data from the sensors. Two questions are important to consider: (1) Where should the sensor reside? (2) How much data should I collect?

Researchers and practitioners typically affix multiple accelerometers to different parts of the body. The body parts are chosen carefully depending on what activity we are monitoring. So for example, activity like "brushing teeth" would be best captured with an accelerometer on the wrist of the user, whereas an activity like "running" would be best captured by observing the lower-body movements. Over the years a few dominant positions have evolved: waist, wrist, chest, feet. There are many wearable computing solutions today [122] that can do an excellent job of monitoring vital physical activities of users with on-body sensors. Typically, bands are used to affix the sensor to the body. On the waist, the sensor is attached to the belt that is worn by most users.

A second strain of work that has evolved over the last few years, investigates the efficacy of learning algorithms to detect activities by using the accelerometer on the smartphone. Body-affixed sensors might not appeal to many users due to its hard constraint to be on-body. On the other hand, the smartphone is a ubiquitous device, carried by most users, and resides in pockets, hands, and handbags. The advantage of performing activity recognition using smartphones is that it imposes no further restrictions to the user. The disadvantage is that due to its flexible body-position, the signals are noisy and may not have high quality. Nevertheless, given the advantages, researchers are investigating several parameters, including how much coverage of physical activities is possible if the smartphone alone is used for semantic activity recognition.

The activity being performed needs to be captured appropriately by the sensor. There are two parameters here: at what sampling frequency should I monitor the activity? For how long? Customized accelerometers may be tuned to several levels of sampling frequency ranging from 5–100 Hertz. Smartphone manufacturers have imposed more strict levels. For example, android has 4 states in which an application can sample the sensor, with a range of sampling frequency for each. The faster the sampling, the higher is the capture of data that resembles the activity. The amount of time is typically guided by the nature of the activity. Simple activities can be easily learnt with multiple samples of a few minutes or seconds. However, complex activities would require longer duration observations. Most works experimenting with smartphones write their own application that resides on the phone and performs data collection based on desired requirements.

Another aspect of data collection is the nature of the environment in which data is collected. There are two dominant styles here : in-lab and in-the-wild. Laboratory data collection

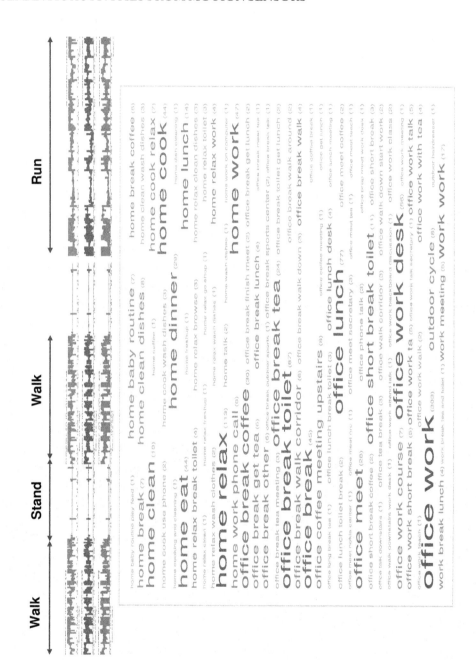

Figure 3.3: Activity learning concepts and a sample tag cloud of activities of daily living (ADL).

experiments select a few users, provide them with sensors, give specific instructions and collect data for research purposes. In-the-wild data collection lets the users go ahead and perform activities in their own immersive environments. In-the-wild data collections have better representation of reality but the data might also have more noise. Researchers thus have to consider how to remove usage-related noise or create learning algorithms without the knowledge of the exact body position of the sensor.

3.2   FEATURE SPACES

Several types of features have been proposed in literature to mine activities from accelerometer streams. In this section, our objective is to provide an organized structure to the features and summarize the dominantly used ones. Let us first define an accelerometer data stream.

Definition 3.1   (Accelerometer Data Stream - $\mathcal{A}$)   Initially, a sequence of data points recording acceleration along 3 axes, i.e., $\mathcal{A} = \langle A_1, A_2, \ldots, A_n \rangle$, where $A_i = (x_i, y_i, z_i, t_i)$ is a tuple with accelerations (x_i, y_i, z_i) and timestamp t_i.

3.2.1   DATA PROCESSING

Due to sensor circuitry, there is always some jitter or noise that is introduced in the signal. Before applying feature extraction, we must apply some standard techniques to remove this error. There are two types of errors: (1) calibration errors and (2) jitter. For example, [169] reports that when a Nokia N95 phone is placed static on the table, and a few seconds of readings are taken, the jitter is around ± 5. It also demonstrates that in theory, x and y axes (the axes along the table surface) should be reporting a value of 0 and the vertical z axis should be having a value of $-g$ where g is 9.81 m/s^2, but in reality they have an offset with the x and y axes report values of around -15. Note that the actual digital output from the sensor is different for different circuitries depending on how the acceleration value is represented digitally.

Signal Smoothing

Let us go through a simple signal smoothing process. Jitters resulting due to the sensor circuitry can be easily removed by using a moving average filter. In this form, we consider a filter of k points. For each point A_i, we replace it with A_i' where A_i' is the average of points around it within a window of k points. Mathematically, this can be represented as:

$$[x_i', y_i', z_i'] = \frac{\sum_{j=i-\frac{k}{2}}^{i+\frac{k}{2}} [x_j, y_j, z_j]}{k}, \tag{3.1}$$

where we are using the previous k points to compute the value of the reading at time t_i. This process reduces the standard deviation of the signal due to the jitters. There can be many variations of this. For example, we can use weighted averages and vary k to obtain various levels of smoothing.

Calibration

Accelerometers are subject to sensor drifts as a result of which they might not provide correct output. For every axis, there might be an offset and a drift of sensitivity. A calibration process is used to find two parameters (offset, scaling factor) to remove the effect of the offset and drift in sensitivity. If this process is not done, the data quality is affected and would impact activity recognition algorithms, working on top of the time series.

Hence, our objective is to explain a simple methodology to understand how to calibrate the sensor. Each of the axes (x,y,z) have their own offset and sensitivity parameters. These axes are orthogonal and independent. We will assume there are no cross-axis sensitivities. Let $a^i = (a_x^i, a_y^i, a_z^i)$ be a sample raw reading from a triaxial ACC at time t_i. Let $g^i = (g_x^i, g_y^i, g_z^i)$ be the calibrated reading in the gravity-scale (g) where $g=9.81 m/s^2$. If the device is static, then it is only subject to g and no other acceleration. Hence, to find the offset and sensitivity factors, the user has to collect data along the $+$ and $-$ directions of each axis, while making the candidate axis strictly point towards the $\vec{g}$ direction, i.e., center of earth. To do this, one can keep the phone in 6 orientations (2x3) on a table that is strictly horizontal to the ground or suspend the phone on a scale or use other means.

Let $a^i = (a_x^i, a_y^i, a_z^i)$ and $a^j = (a_x^j, a_y^j, a_z^j)$ are two readings obtained for the $+$ and $-$ directions of X-axis. We can calculate $offset_x$ and $scale_x$ as follows:

$$offset_x = \frac{a_x^i + a_x^j}{2}$$

$$scale_x = \frac{\mid a_x^i --- a_x^j \mid}{2}. \tag{3.2}$$

It could be a challenge to take such sensitive readings and errors might creep in if the orientations are not correct during the calibration process. [170] describes an improved process where it can tolerate some manual errors. Readers are encouraged to read the methodology to obtain further details. In a nutshell, this method takes advantage of the fact that the mean of accelerometer readings along each axis over a long period of time can produce a good estimate of $\vec{g}$ direction.

3.2.2 FEATURE CLASSES—AN OVERVIEW

Let us assume that we have access to a data stream $\mathcal{A}$ of accelerometer readings. This data stream is first split into *frames* of fixed intervals. The frame size T_f may be of the order of seconds. The frame size may also be such that each frame contains a fixed number of readings. Typically, when an accelerometer is sampled at a certain sampling frequency, the exact number of readings per second has a slight standard deviation but largely stays uniform with the sampling frequency as the mean. For example, if the frequency is 5 Hertz, then we are recording 5 readings per second on an average. Let us call the jth frame as A^j. Each frame A^j in turn consists of a series of acceleration readings a_x, a_y, a_z along the three dimensions of the ACC.

Next, we will discuss the different ways in which features can be extracted from the data frames. To summarize the body of work, we observe that the features can be organized in a matrix style: (orientation X domain). The features can be either orientation-independent or orientation-sensitive. The features can be from the time domain or the frequency domain of the time series of readings.

Orientation-sensitive feature extraction

Note that the orientation of the phone changes with the usage, placement of the device. As a result, the axes (as described before) also point in different directions, compared to a fixed reference frame. Orientation-sensitive features are all statistical features whose values would alter depending on the orientation of the device. For example, if we consider A_x^j to be the mean of x-axis readings of all points in the jth frame, then this value would alter depending on the exact direction $\vec{x}$ is pointing to.

Orientation-independent feature extraction

The advantage of the orientation-sensitive strategy of feature extraction is that the data is of high quality and as a result the classifiers can learn the activities well. The problem with this strategy however, is that it might treat the same activity, performed with two different device orientations as two different activities. This is because the feature values would be varying based on the orientation. For example, if we perform *walk* with the smartphone in two body positions—clipped to the belt and inside the pant pocket, the orientation-sensitive feature values would vary greatly due to the changed orientations. To address this problem, there are two possibilities. One possibility is we use statistical features that are not affected by orientation. For example, the mean of the readings in a frame $- \sqrt{a_x^2 + a_y^2 + a_z^2}$ does not have a direction and hence is an orientation-independent feature. However, magnitude-based features lead to information loss.

To avoid the impact of orientation of the phone on the activity signature, the signal can be projected to a fixed reference frame. In [112], Mizell describes a method for transforming the signal to a fixed reference frame where one axis points towards the gravity vector ($\vec{g}$) and the other is on the plane perpendicular to $\vec{g}$. All MEMS accelerometers measure the gravitational acceleration ($\vec{g}$) as well as the dynamic accelerations caused by the phone user. They find that the mean of accelerometer readings along each axis over a long period of time can produce a good estimate of the gravity direction.

The method works as follows.

- For a given, but arbitrary, orientation of the device, calculate the impact of $\vec{g}$ on each axis by measuring the average of the readings in each axis separately, taken over a time interval, typically at least a few seconds. Let this be called the vertical acceleration $v = (v_x, v_y, v_z)$, where v_* are the measurements on respective axes.

- Let $a = (a_x, a_y, a_z)$ be the measurements for a given point in the sampled frame. Find $d = (a_x - - - v_x, a_y - - - v_y, a_z - - - v_z)$ which would represent the impact caused by the user's movement. Let this be called the dynamic component vector.

- Using dot product of these two vectors, we can then find the projection of d on to the reference frame. $p = \left(\frac{d.v}{v.v}\right) v$, where p is the vertical component of d.

- Since a 3D vector is the sum of its vertical and horizontal components, we can compute the horizontal component of d by vector subtraction, as $h = d - - - p$. However, unlike p, we do not know the direction of h. All we know is that it is on the plane perpendicular to the gravity vector $\vec{g}$.

Time and Frequency Domain Features

Time domain features are all features that may be computed from the signal, represented as a time series of readings. Researchers have also converted the signal into the frequency domain by performing fourier transforms on the readings and then computing several frequency domain features.

Below, we provide a listing of features that have been used repeatedly in the literature. Let us assume that a frame T_f contains n points $[x_i, y_i, z_i](i = 1 \ldots n)$.

- **Axis-specific means ($AVG(\mu)$), variances (VAR).** For each frame, compute the mean and variance for each axis separately. For example, $\mu_x = \sum_{i=1}^{n} x_i$. Similarly, $VAR(x) = \frac{1}{n} \sum_{i=1}^{n} (x_i - \mu_x)^2$.

- **Magnitude (MAG).** Magnitude of the frame is the average of magnitudes of all the point vectors. For example, $MAG(x, y, z) = \frac{1}{n} \sum_{i=1}^{n} (\sqrt{x_i^2 + y_i^2 + z_i^2})$.

- **Correlations ($CORR$).** Linear associations between two axes in a frame may be represented as a feature. The correlation coefficients between two axes are computed using the standard formula $corr(xy) = \frac{cov(xy)}{\sigma_x, \sigma_y}$, where σ_x, σ_y are the standard deviations observed in the x and y values in the frame.

- **Signal Magnitude Area (SMA).** Signal Magnitude Area can be calculated using the time series of the signal, as the area under the magnitude of all the three axes. For a discrete space, it may be $\frac{1}{n} \sum_{i=1}^{n} (|x_i| + |y_i| + |z_i|)$, where $|x|$ represents the magnitude of the $\vec{x}$, and n=number of points. This feature does not have a direction and hence can be applied for activity recognition when the device orientation is non-static.

- **Zero Crossing Rate (ZCR).** Zero crossing rate is a simple feature that measures the number of samples in a frame that crosses the zero reference line. If the ZCR is high, it is an indicator that the signal might have high noise. It can also be substituted by mean crossing rate, replacing the zero reference line with the mean of the signal.

- **Fast Fourier Transform Features (FFT Features)** [mean, correlations, spectral roll-off, spectral centroid, spectral flux]. Fourier transform converts the time series to its frequency domain. Features extracted from this frequency domain can be used to detect activities. Typically, activities that have periodic peaks (e.g., walk) can be well detected using the FFT features. By counting the number of peaks in a signal, these features can also be used to compute foot step counts for example. Today there are many opensource tools and algorithms that compute the Fourier transforms, and features on top of it. Researchers have also tried out features that try to capture the shape of the frequency spectrum using different means. For example, [98] reports the results of applying a logarithmically spaced filter bands to measure ADLs and reports that this performs better than the use of the default linear spacing of bands.

3.2.3 FEATURE COMPUTATION AND ENERGY

The compute time of features varies based on the complexity involved in computation. Typically, the time-domain features are faster to compute than the frequency domain features. The sampling rate at which the accelerometer data is collected impacts the quality of raw observations as well as volume of data being collected. Hence, this too plays a role in the compute time. The computation platform (server or a mobile device) is yet another factor impacting the feature computation.

The vast majority of literature on accelerometer-driven activity sensing focuses on either (a) the identification and use of progressively more sophisticated *features* for more accurate activity classification [170] or (b) the use of multiple body-worn accelerometers for improving the accuracy of activity recognition [66, 139]. Studies (e.g., [91, 111]) on mobile sensing have, however, observed that continuous activity recognition (applying on-board data processing over accelerometer and other sensor data streams) can rapidly drain the power on mobile devices, often dropping the operational lifetime to 3-4 h.

Early research on the use of individual or multiple body-mounted accelerometers for locomotive activity sensing focused on understanding the various factors affecting the recognition accuracy, e.g., [54] demonstrated that different activities exhibited differing levels of classification accuracy, depending on the on-body placement of the accelerometers.

The study on the eWatch platform in [91] discusses the impact of time and frequency domain features towards power consumption. [9] studied the classification accuracy based on multiple body-worn accelerometers, as a function of a more complex set of both time and frequency domain features (such as mean, energy, and entropy).

Most of these results do not directly apply to a commercial smartphone device. This is primarily because the smartphone has a multi-purpose functionality and the background energy consumption is quite different from that of a custom sensor performing a focused and optimized function on the hardware. Along these lines, the use of a secondary sensing "co-processor" has been shown [104, 132] to provide highly energy-efficient activity recognition on mobile devices; such hardware-based optimizations are helpful to scale the computational logic of activity sensing.

More recently, adaptive online techniques have been proposed to reduce the energy over-heads specifically associated with personal mobile devices. A typical approach is called smart duty-cycling. In this approach [160], power-hungry sensors are turned on only if less energy-hungry sensors detect a significant event. The Kobe toolkit [34] focuses on balancing the accuracy and energy cost of activity classification algorithms on mobile devices, by dynamically adjusting the pipeline of sampling frequency, feature extraction and machine learning components. The runtime adaptation focuses on adjusting where the computation is executed (device. vs. cloud), in response to system changes (such as changes in phone battery levels or foreground process-ing load). Similarly, the SociableSense system [136] applies reinforcement learning techniques to adjust the duty cycle of multiple sensors and uses a multi-criteria decision theory to distribute the computational tasks between a mobile device and the cloud. SociableSense makes a binary decision ("should I sense or not"), based on information in prior sensing cycles.

Our recent work [168] progresses the state-of-the-art by focusing on two independent parameters of accelerometer-based activity recognition: (a) the sensor sampling frequency and (b) the set and classes of features used in the activity classification process. Investigations (e.g., [83]) have established that these two parameters jointly influence a tradeoff between two impor-tant and mutually-conflicting objectives:

1) **increase classification accuracy.** increase in sampling frequency and a richer set of features both result in improved activity classification accuracy; and

2) **reduce energy overheads.** conversely, reducing the sampling frequency, duty cycle and/or the data processing to compute feature vectors should result in a lower energy overhead.

This work reports that different activities indeed differ in how their classification accuracy varies with changes to the sampling frequency and the set of classification features used. Taken together, these studies provide an important insight: the most judicious *combination* of (sampling frequency, classification feature) for activity recognition is *different for each distinct activity*. More details of the methodology to jointly adjust the two parameters (sampling frequency, classification features) can be found in [168].

3.3   ACTIVITY LEARNING TECHNIQUES

In this section, we discuss the key techniques to learn activities from smartphone accelerometer streams. Once the raw accelerometer data stream has been divided into frames (T_f) and trans-formed into a vector of features, these feature vectors are used to learn the models of the activities they represent.

As discussed before, human activities can be grouped into two categories: (1) micro-activity and (2) complex activity. Let us first create some principal definitions:

Definition 3.2 **(Micro-Activities $\mathcal{MA}$)** A set of M distinct micro-activities $\mathcal{MA} = \{MA_1, \ldots, MA_M\}$, where each element of $\mathcal{MA}$ corresponds to a pre-defined locomotive state of the individual.

Definition 3.3 **(Complex Activities $\mathcal{CA}$)** A set of H distinct activities $\mathcal{CA} = \{CA_1, \ldots, CA_H\}$, where each element of $\mathcal{CA}$ corresponds to a user-defined tag for an activity of the individual and where each $\mathcal{CA}$ consists of one or more MAs.

Figure 3.4 demonstrates a hierarchy of complex activities obtained from a survey [163]. The hierarchy reflects activity tags at different levels, encompassing the activities at the lower levels.

3.3.1   LEARNING MODELS

Let us consider a stream of sensor readings coming from the ACC, affixed to a certain user. At any point of time (t), there are two questions that arise. $Q1$: What is the current activity ($a_i \in A$) being performed, where A is the vocabulary of all activities? $Q2$: At what time ($t' \leq t$) did this activity start?

The typical methodology applied towards learning the activity is to take some data samples from the sensors and learn the data patterns exhibited by the activities. Depending on where the sensor is placed, the data patterns reflect the limb movements dominantly. The more observations we have, the better is our learning capacity. For example, an ACC fixed only on the wrist of a user cannot distinguish between *sitting* and *standing*, unless the arm shows some significant movement difference. However, a sensor near the waist of the user can carry meaningful patterns to distinguish between these two activities.

Over the years, researchers have applied several machine learning techniques to learn these data patterns. They may be categorized into two groups: supervised and unsupervised. The key aspect that makes these methods different is the assumptions they make. In supervised methods, the models of patterns are learnt from training data sets annotated with tags which contain information of the actual activity. Let us call this annotated training data as "ground truth." In unsupervised methods, the patterns can be learnt without relying on the ground truth. The primary focus is to learn the key discriminative clusters of patterns exhibited by a user or a group.

For detecting activity start (and end) times, algorithms applied are called "change point detection" algorithms. They primarily work on trying to identify when a certain pattern finishes and the next starts, in a data stream.

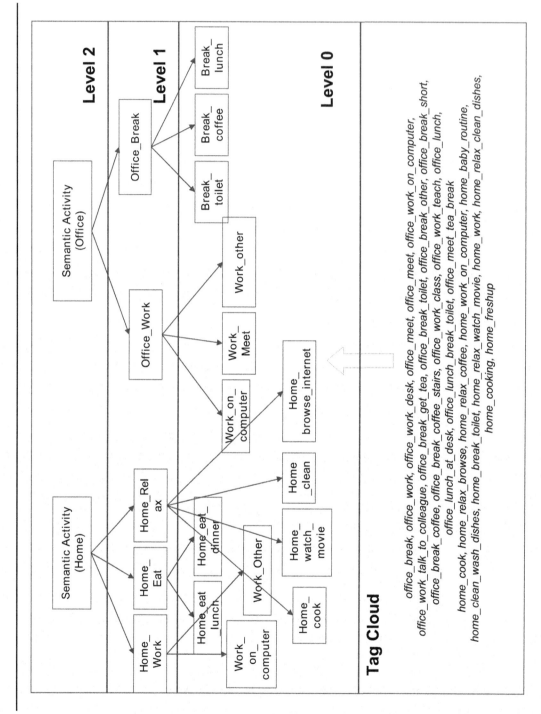

Figure 3.4: User tag cloud and hierarchy of complex activities.

Supervised Learning Models

The body of work on supervised learning models can be primarily divided into two groups: (1) memory-less learning and (2) learning with history. Memory-less learning, as the name suggests does not take any prior history into consideration. The prediction is based solely on the current observation coming from the accelerometer frame. In the second approach, the past sequence of predictions or actual labeled data is used to learn additional patterns of what activity might precede or follow what activity. Accordingly, these patterns are used for predicting the current activity.

Both techniques may be used separately or in conjunction for real-world data sets. Next, we describe briefly a few standard learning techniques of these types. Note that these learning techniques are all well known in machine learning community and interested readers are encouraged to learn more about them from focused literature that describes these in further detail.

- **Decision Trees.** Also known as Classification Trees, this methodology learns a tree structure which predicts the value of a dependent attribute, given the value of the other independent ones. There are well-known models to learn a decision tree that best fits the training data. For example, C 4.5 is a well-known decision tree implemented within the WEKA framework. Decision trees while being widely adopted, have their pitfalls. It can easily over-fit the training data, in particular for disbalanced data sets, and they also try to fit the noise. Decision trees are also known to have brittle boundaries, particularly when data is numeric.

- **Bayesian Models.** These models predicts a class based on the likelihood of a set of observed features to be present for that class. They make a simple (naive) assumption that presence or absence of a feature for a class, is independent of the presence or absence of another feature for that class. The class that has the maximum likelihood of a set of observed features to be present is what is predicted.

- **Support Vector Machines (SVM).** Support Vector Machines work with the fundamental idea of trying to find a hyper plane between a set of points representing classes such that it maximizes the separation between the classes. This is typically well suited when the classification problem has high dimensionality, i.e., a lot of features represent each class.

- **Adaptive Boosting.** This is a form of linear classification introduced in 1995 [57] that constructs a strong classifier as a linear combination of weak classifiers. A weak classifier is one that has two properties of its class prediction logic: (1) it is simple or easily computable and (2) prediction logic only partially correlates with the true classes. The algorithm works by incrementally trying to learn more from the data sets that are falsely classified by the previous weak classifiers.

- **K-Nearest Neighbors.** This classifier is a simple one that works on the principle of labeling a data set with the same label of a known data set that is *nearest* to the unlabeled one, using some distance function. Although simple in nature, it is sensitive to noise in the training data.

The models that learn from history of observations may further improve the confidence of predicting what activity is occurring at present. For example, hidden Markov model (HMM) is a classical statistical signal model in which the system being modeled is assumed to be a Markov process with unobserved state. We may consider a temporal sequence of accelerometer vectors as observations, and predict the hidden sequence of activities that are most likely causing the current observations. Discriminative models such as conditional random fields (CRF), latent Dirichlet allocation (LDA) are commonly applied over observations to learn about activity distributions. The focus of this chapter remains on learning the methods that one may apply, even if there is no history present. In the case study section, we would cover in some more details how we may use LDA to extract features for long-running complex activities.

Unsupervised Learning Models

There are some drawbacks on learning using supervised approaches.

- **Training Burden.** We need to request the users to spend non-negligible time tagging their activities. This is an onerous task for the user. In addition, the user often needs to be shadowed, with a diary, in order to collect such data. Moreover, the user has to avoid using the smartphone for regular purposes during this data collection, in order to collect representative samples of the activities. The more fine-grained the tagging requirement, the higher cognitive burden the user has. For example, tagging all micro-activities performed by an user in her day is a challenging task.

- **Inability to capture complete reality.** Due the above cognitive hindrance of collecting tagged data, a survey-driven method captures only a limited set of activities performed by the user. Clearly, in real-life, while performing complex activities, users may perform several other diverse micro-activities. For example, walk itself can have several variations, like walk slowly, walk briskly, walk with a strut, etc. A user is expected to perform many kinds of such numerous real-life micro-activities. It is impractical and difficult to capture all of them through a supervised training set.

Unsupervised clustering techniques may be used to learn activities from untagged data streams. For example, we can perform a k-means clustering on an accelerometer data stream. In order to do this, the raw data in each accelerometer frame is represented as a vector of statistical features. This approach provides a data-driven mechanism to group frames that represent similar (but unknown) micro-activities, thus eliminates the training burden. Secondly, they represent reality, i.e., personal patterns are more likely to be captured using clusters.

We may vary the number of clusters (K) to determine the best performing clustering configuration. Well-known clustering metrics like DBI (the Davies-Bouldin index) K [73] may be applied to evaluate the clustering performance.

3.3.2 CHOICE OF LEARNING MODELS

The choice of which classifier to use is a difficult question to answer. It depends on a few factors, which we would like to enumerate.

First, one needs to consider where the classification task should be performed. If the classification learning and prediction is performed on the mobile device, one must pay particular attention to the computational cost of learning. For example, decision trees with arbitrary long branches can be difficult to execute on a device.

Second, for training, one needs to consider the nature of the training data and how it was collected. Note that as discussed earlier, training data in accelerometer-based activity learning can be collected in different ways, ranging from collecting data in the lab by carefully controlled experiments, to collecting data in the "wild." Mislabeled data can be a source of poor classification performance. Similarly, correctly labeled data but containing usage-induced and user-specific variations can make it difficult for a classifier to learn its models.

Third, one needs to consider the final application. Is this classifier meant to be used on historically logged data records, primarily for mining purposes, or is it meant to be used for real-time classification of incoming streams? The computational complexity of predicting the class in real-time plays an important role for real-time classification.

Forth, one needs to evaluate the amount of data to be classified. Accelerometer data streams are a form of time series. Micro-activities would require short samples of the data, while complex activities would require larger chunks of the stream.

Finally, the decision output may be absolute or fuzzy, i.e., probabilistic. This is also referred to as "soft" or "hard" output. This can be a consideration for some applications.

3.3.3 TESTING

For testing, classifier models are first trained using a part of the data, while keeping the rest for testing. The following modes of testing are common in activity recognition:

- **N-fold cross validation.** In this technique, a data set is randomly divided into N sets. N-1 sets are used for training the classifier and the remaining one set is used for testing. This process is repeated N times with each of the N sets used exactly once for testing. The results from the N runs are averaged to produce the mean and the standard deviations of key parameters like precision and recall.

- **Random sub-sampling validation.** In this technique, the data set is randomly split into training and test sets. Subsequently, after training and testing with the current split, the process is repeated with another random split. The results of N runs are averaged. The advantage of this method is that the training data size does not depend on the number of runs, however, the results might vary from one set of runs to the next, as a data item may be picked up multiple times or not at all.

- **Leave one out validation.** In this method, one sample is kept out and the rest is used for training. This methodology is useful when we would like to test how models trained with data of N users, apply towards the $N + 1^{th}$ user.

Next, let us go over two case studies with some real-world data sets, and discuss the salient aspects for learning activities from smartphone accelerometer streams.

3.4 CASE STUDY: MICRO ACTIVITY (MA) LEARNING

It is important in activity recognition research to describe specifically how the activity data is collected, for a reader to understand the results.

3.4.1 DATA COLLECTION

To evaluate the MA classification performance, we recruited 5 participants, who were each provided a Nokia N95 phone with embedded Python scripts that sampled the accelerometer sensor at 30 Hz for the duration of the study. For ease of explanation, let us consider a set of 7 ($M = 7$) micro-activities: {"sit," "sit relax," "walk," "slow walk," "bursty move," "stand," "using stairs"}. While most MA elements are self-descriptive, the non-obvious ones are described in Table 3.1. Each user was asked to perform these ($M = 7$) 7 MAs consecutively for ~7–10 min each, resulting in a per-subject study duration of ~50–60 min.

Table 3.1: Descriptions of some non-obvious micro activity labels

Name of MA Label	Exemplary Description of Activity
sitRelax	all types of relaxed sitting (e.g., shaking legs, stretching)
slowWalk	walk at a slow pace, walk inside office rooms
burstyMove	jerky movements (e.g., get up from chair, movements inside kitchen), varies across users

Our data collection is in naturalized setting compared to traditional wearable computing with fixed on-body sensors. We first collected data with the phone placed only on the user's *preferred* body position, i.e., where they predominantly carry their phones. However, the phone can be placed on different body positions with varying orientations. Therefore, we additionally perform another data collection on some users. In this study, the users performed each MA while changing the phone's on-body position among: {*Shirt Pocket, Pants Front Pocket, Pants Back Pocket*}, and while randomly altering the phone's orientation in each of these positions. This emulates the real-life model in which phones are used and placed in different body parts for long-term usage.

We are thus able to study the classification accuracy under the following naturalistic conditions:

a) when the subject carries the phone in the position (which we know a-priori) that he/she most prefers while performing daily chores; and

b) when the exact on-body position and orientation of the phone is unknown and subject to random changes.

3.4.2   FEATURE VECTOR REPRESENTATION

We first followed the techniques of signal smoothing and calibration as described in the section on data processing (Section 3.2.1). Our next task is to convert the raw data stream into a set of feature vectors, representing the MAs. As discussed before, to do this, we divide the data into frames, each frame containing data for 5 s. Our objective in this chapter is to investigate a broad set of statistical features and classification algorithms that operate on the accelerometer data ($\mathcal{A}$), and empirically understand how much micro-activity classification accuracy can be achieved under naturalistic conditions.

Since the device can be in different body positions with varying orientations, recent literature (e.g., [33, 106]) stresses the importance of understanding the performance of these features under naturalistic conditions. Following the feature classes introduced earlier, we summarize the groups of features we have.

1) **Orientation-Dependent Features.** These features are computed separately on each of the three axes, x, y, z (e.g., mean values $\{\bar{x}_i, \bar{y}_i, \bar{z}_i\}$ of the raw data in frame F_i). In general, we can expect orientation-dependent features to be useful when the phone's orientation relative to the earth's horizontal remains constant.

2) **Orientation-Independent Features.** These are either: (1) an orientation-insensitive combination of the readings from the axes (e.g., the variance of the acceleration magnitude $\sqrt{x^2 + y^2 + z^2}$); or (2) associated with acceleration values *projected to a reference frame, e.g., the ground.* We expect orientation-independent features to be more robust to variations in a phone's orientation/on-body location, but offer lesser resolution than their orientation-dependent counterparts.

We explore various *combinations* of these two feature classes, and evaluate their ability to classify the locomotive activities, which have varying degrees of similarity (e.g., "walk" and "loiter"). We consider a feature vector, consisting of both time and frequency domain features from: (1) the 3D axis of the phone, referred to as 3D-features; and (2) A projection of the readings on the gravity direction ($\vec{p}$) and the plane perpendicular to gravity($\vec{h}$), which makes it orientation-independent (referred to as 2D-features). We use the fact that the mean of accelerometer readings, computed over a long time period gives an estimate of g [106], to project the raw signal to this "2D" reference frame. For the frequency domain, the features are computed by first transforming the (x_i, y_i, z_i) segment into a 250-point FFT vector [139]. Finally, a total of $\sim$70 features are used per frame (F_i). Table 3.2 summarizes the feature types, where the last column (OI—Orientation Independent) indicates whether the feature is orientation-independent or not.

Table 3.2: Features used for micro-activity classification

	Name	*Definition*	*O*
Feature Components	calibrated 3 axis data (3D)	(x_i, y_i, z_i)	N
	projected 2D (Vertical) $[\vec{p}]$	$\vec{p} = \frac{d \cdot \vec{v}}{\vec{v} \cdot \vec{v}} \cdot \vec{v}$, where $v = \langle \bar{x}, \bar{y}, \bar{z} \rangle$ (the mean of x,y,z) and $\vec{d} = \langle x - \bar{x}, y - \bar{y}, z - \bar{z} \rangle$	Y
	projected 2D (Horizontal) $[\vec{h}]$	$\vec{d} - \vec{p}$	Y
	projected 2D (Magnitude) $[mag]$	$|\vec{h}|, |\vec{p}|, corr(|\vec{h}|, |\vec{p}|)$	Y
Time Domain Features	Mean	$AVG(\sum x_i); AVG(\sum y_i); AVG(\sum z_i)$	N
	Variance	$VAR(\sum x_i); VAR(\sum y_i); VAR(\sum z_i)$	N
	Mean-Magnitude	$AVG(\sqrt{x_i^2 + y_i^2 + z_i^2})$	Y
	Magnitude-Mean	$\sqrt{\bar{x}^2 + \bar{y}^2 + \bar{z}^2}$	Y
	Two-Axis Correlation	$corr(xy) = \frac{cov(xy)}{\sigma_x, \sigma_y}$; similarly $corr(yz), corr(xz)$	N
	Signal-Magnitude Area (SMA)	$\frac{1}{n} \sum_{i=1}^{n} (|x_i| + |y_i| + |z_i|)$	Y
Frequency Domain Features	FFT Magnitude	$m_j^{(x)} = |a_j + b_j i|$; similarly, $m_j^{(y)}, m_j^{(z)}$	N
	FFT Energy	$\frac{\sum_{j=1}^{N} (m_j^2)}{N}$, for x,y,z respectively	N
	FFT Entropy	$-\frac{\sum_{j=1}^{n} (p * log(p))}{n}$, for x,y,z respectively, where p is Normalized histogram count of FFT component magnitudes	N

3.4.3   RESULTS OF MA SUPERVISED LEARNING

We present the results using a *10-fold cross validation approach*, where each run utilizes 90% of the collected data as the *training* set, and the resulting classifier model is applied to the remaining 10% *test* data.

Figure 3.5 plots the classification accuracy with various choices of 3D+2D feature vectors for five users, with the phone located in each individual user's preferred on-body position (as indicated, this position varies by user). We test many classifiers[3] on these feature choices. Furthermore, we also applied correlation-based feature selection measure [145] to selectively boost good features. In the figure, the plot is the average accuracy over all of the classifiers and includes the standard deviation of the accuracy values. The plots correspond to a frame duration of $T_f = 5$ secs. While LibSVM and Adaboost achieved better results, in general, than others, there was no universal winner among the classifiers. The plot shows that the classification accuracy is uniformly high across all users, dropping to no lower than 88–89% in the worst case. We also observe that

---

[3]*Decision tree –J48 Naive Bayes, Bayesian network, LibSVM* and *Adaptive Boost* (Adaboost) using J48 as the weak learner

correlation-based feature reduction, applied on a combined set of $3D_{all}+2D_{all}$ features, provides marginally better performance.

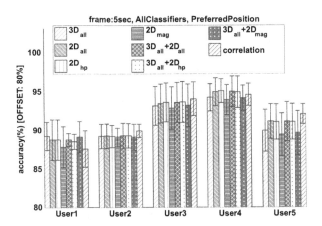

Figure 3.5: Micro-activity classification accuracy across all 5 users ($T_f = 5s$) with the phone in their preferred positions —{front pocket in the pants, back pocket in the pants, shirt pocket in the chest}. The plot shows various combinations of "3D" & "2D" features (ref. Table 3.2). $3D_{all}$ (or $2D_{all}$) implies the use of all time and frequency domain 3D-features (or 2D-features); $2D_{hp}$ refers to features computed on the projected orientation-independent frames—both gravity ($\vec{p}$) and its plane perpendicular ($\vec{h}$) ; $2D_{mag}$ refers to features computed over magnitudes of $\vec{h}$ and $\vec{p}$; and "correlation" refers to the feature selection technique we used.

Figure 3.6 plots the classification accuracy achieved on $User1$ when the orientation of the phone (within each on-body position) changes dynamically. The last column reflects the results when the on-body position of the phone is assumed to be "unknown." In this case, the classification is performed and the accuracy is evaluated on the *combined* data of all 3 positions. The figure illustrates the following points: (1) the MA classification accuracy is higher when the phone is placed in the lower part of the body (an observation also made with multiple body-worn sensors [9][66]); (2) the choice of feature classes result in performance differences of ∼5%; and (3) the classification accuracy for the "unknown" case, which best reflects naturalistic usage conditions is reasonably high ∼90%;

Figure 3.7 plots the "confusion matrix" for the "unknown" case– the results show that our combination of features is able, in most cases, to disambiguate correctly among the MAs.

3.5 CASE STUDY: COMPLEX ACTIVITY LEARNING

Our objective in this section is to discuss complex activity learning, via a specific case study. First, we describe the data collection.

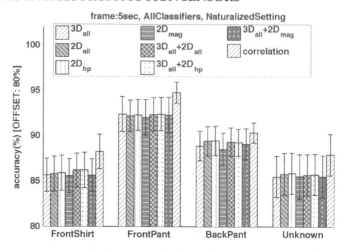

Figure 3.6: MA classification accuracy for *User1* with naturalistic (varying) phone orientations ($T_f = 5$ secs) and with unknown body positions. 'FrontShirt'=shirt pocket in the chest, 'FrontPants'=front pocket in the pants, 'BackPants'=back pocket in the pants. 'Unknown'=body position is mixed and not given.

$$\mathcal{T} = \begin{pmatrix} & bm & sr & s & sw & r & t & w \\ burstyMove(bm) & 50 & 1 & 0 & 12 & 1 & 0 & 2 \\ sitRelax(sr) & 3 & 47 & 0 & 2 & 0 & 14 & 0 \\ sit(s) & 0 & 2 & 19 & 0 & 0 & 9 & 0 \\ slowWalk(sw) & 6 & 3 & 0 & 54 & 1 & 0 & 0 \\ stairs(r) & 3 & 0 & 0 & 0 & 54 & 0 & 6 \\ stand(t) & 1 & 1 & 0 & 0 & 0 & 64 & 0 \\ walk(w) & 3 & 0 & 0 & 8 & 2 & 0 & 53 \end{pmatrix}$$

Figure 3.7: Confusion matrix for MA classification of *User*1. An entry in the ith row & j^{th} column denotes the number of instances where the classifier labeled the activity as MA_i, while the "ground truth" was MA_j.

3.5.1 USER RECRUITMENT AND DATA COLLECTION

Unlike MA data collection, in order to effectively study the power of the ACC to classify Complex Activities (CA) of daily living, we have to collect significantly more data across several days. During the data collection campaign, 5 users volunteered to carry a Nokia N95 phone, loaded with an application that sampled the accelerometer at 30 Hz, continuously 24×7. Four users used it as their primary cellphone for the duration of our data collection. Users were instructed to tag their activities in a separate diary while they conducted chores. To limit the number of activities, we monitored CAs only in their home and office locations. The only specific instructions given were that the users should carry the phone with them, in their preferred way, while they were tagging activities, and that they should charge the phone only when they were not tagging themselves (typically in the night during sleeping). This longitudinal data was gathered over a span of 8 weeks on working days.

The user tagging process and principle followed was *unconstrained*—users had their own discretion on what activities to tag. The users were provided an initial idea on what constituted a complex activity (e.g., work, break, lunch). Although not mandatory, users could provide additional detail for each activity (e.g., break_coffee, break_toilet, work_at_desk). Each user recorded the tag tuples: [*activity start time, activity tag*]. As the activities were sequential, the end time of an activity was derived from the start time of the next tag. The last activity performed in a sequence, on a certain day at a certain location had an explicit end time registered by the user. In the aforementioned Figure 3.3, we have already presented the tag cloud from our data collection campaign. There are 1284 tags in total, with 177 unique tags.

3.5.2   DATA PROCESSING & SANITIZATION

The data was cleaned by applying a per-user manual process of normalization and information summarization. (1) Semantically equivalent tags (e.g., office_meet, office_meeting) were converted to a standard notation. (2) Tags having additional context were collapsed to the corresponding root tag (for instance, office_meet_colleague → office_meet), unless the activity occurred very frequently, and vice versa, e.g., the activity office_break_toilet was separated from office_break for some users. Infrequent tags were subsequently removed from further investigation (e.g., home_freshenup). In total, we obtained 152 days of data, with each day containing between 4-15 tags/person. Table 3.3 provides the person-specific summary of the collected data. The final, cleaned data contains records of a total of 1102 complex activities across all users.

Table 3.3: Summary of complex activity dataset

	User1	User2	User3	User4	User5
# of Days	27	31	39	32	23
# of unique HAs	30	64	25	41	65
# of activities	194	215	372	167	228
# of activities used	186	203	356	165	192

3.5.3   FEATURES FOR COMPLEX ACTIVITIES

We now need to derive the features, based on this sequence of frame labels. We present a few feature extraction strategies next. Note that our focus is to only discuss the features from the data

Table 3.4: Examples showing activities collected (right column) and corresponding normalized tags (left column)

HA Label	Examples of User Tags
O_work	office_work, work_work, office_work_TA, office_work_check_printer
O_break	office_break, office_break_walk around office_break_talk,
O_coffee	office_coffee_break, office_break_tea, office_short_break_coffee
O_toilet	office_break_toilet, work_break_toilet, office_short_break_toilet
O_meet	office_meet, office_meet_lab, office_meeting, office_meet_NRC
O_lunch	office_lunch, work_lunch, office_lunch_desk, office_break_lunch
H_work	home_work, home_work_move, home_work_on_computer
H_relax	home_relax, home_relax_freshen_up, home_relax_movie
H_break	home_break, home_break_shopping, home_break_coffee
H_cook	home_cook, home_cooking, home_clean_dishes, home_wash_dishes
H_eat	home_eat, home_lunch, home_dinner, home_eat_adults_with_movie
H_baby	home_baby_routine, home_baby_routine_eat_with_baby

stream. Additional contextual features like time-of-the-day may also be used to capture temporal regularities of CAs and to improve classification accuracy.

Statistical Features

Here, we take the data chunk representing each CA and extract the standard set of statistical features as discussed earlier. Thereafter, a CA is represented as a vector of feature values, extracted from the raw ACC data of the CA instance.

Set and Sequence Features

These features are extracted after the raw stream is converted into a sequence of MAs using MA classification techniques on top of the raw data. Note that we may apply supervised or unsupervised (clustering) techniques to convert the raw stream into the MA sequence.

   We utilize the illustrative example in Table 3.5 to explain these feature extraction strategies. The table shows two CA instances (i.e., CA_1 - *Office_Break* and CA_2 - *Office_Lunch*) from training data, with a simple set of MAs, noted as "walk (w)," "sit (s)," "stand (t)."

Order-Oblivious (OO)/Set Feature Extraction. Given the MA sequence of a CA instance, the *Order-Oblivious (OO)* approach creates an M-dimensional feature vector (M = the number of MAs), where the ith element of the vector denotes the number of MAs of type MA_i. The feature vector thus captures the duration (as T_f is a constant) of each specific MA in the given CA instance. For example, in the first column of Table 3.5, the first CA instance's MS sequence $MS_1^{(1)}$ = [t t t t t t w w w w t] has 4 'walk' MAs, 0 "sit" MAs, and 7 "stand" MAs. The corresponding feature-vector for $MS_1^{(1)}$ is [4, 0, 7].

Table 3.5: Running example of set & sequential features

Col. (Column) 1	Col. 2	Col. 3 SA-TD Patterns Subseq (size: 3, θ≥0.6)			Col. 4 SA-TD Features	Col. 5	Col. 6 SA-TP Patterns Subseq (size: 3, θ≥0.6)			Col. 7
MA Streams of 2 Types / HA₁: Office_break / HA₂: Office_lunch / [w:walk, s:sit, t:stand]	OO Features / [w, s, t]	sub_c	cov	supp	SA-TD Features / [w, s, t, ttw, tww, wwt, tts, tss, sst]	T-P Seq	sub_c	cov	supp	SA-TP Features / [w, s, t, twt, tst]
HA₁ MS₁⁽¹⁾: [tttttwwwwt]	[4, 0, 7]	[ttt]	1	1/3	[4,0,7,1,1,1,0,0,0]	[twt]	[twt]	2	2/3	[4,0,7,1,0]
MS₂⁽¹⁾: [tww]	[2, 0, 1]	[ttw] / [tww]	2 / 3	2/3 / 3/3	[2,0,1,0,1,0,0,0,0]	[tw]				[2,0,1,0,0]
MS₃⁽¹⁾: [ttwwtt]	[2, 0, 4]	[www] / [wwt] / [wtt]	1 / 2 / 1	1/3 / 2/3 / 1/3	[2,0,4,1,1,1,0,0,0]	[twt]				[2,0,4,1,0]
HA₂ MS₄⁽²⁾: [ttsssst]	[0, 5, 3]	[tts] / [tss] / [sss]	2 / 2 / 1	2/2 / 2/2 / 1/2	[0,5,3,0,0,0,1,1,1]	[tst]	[tst] / [wts]	2 / 1	2/2 / 1/2	[0,5,3,0,1]
MS₅⁽²⁾: [wwttsst]	[2, 2, 3]	[sst] / [wwt] / [wtt]	2 / 1 / 1	2/2 / 1/2 / 1/2	[2,2,3,0,0,1,1,1,1]	[wtst]				[2,2,3,0,1]

Sequence-Aware (SA) Feature Extraction. This approach extracts additional features that capture the *order* (or sequence) in which the various MAs occur within a specific CA instance. This approach should improve discriminatory capability of the resulting features, compared to the *OO* approach which does not utilize such knowledge. However, it comes at the expense of higher dimensionality of the feature vector. We consider two pattern mining-based techniques to learn such key discriminatory features from the underlying traces: **SA-TD**, a *duration-preserving* strategy and **SA-TP**, and a *transition-preserving* strategy. To explain them, we first define a few terms.

Let $M_i = [MS_1^{(i)}, \ldots MS_l^{(i)}]$ be the set of Micro-Activity sequences associated with the l different instances of CA_i. For example, $MS_1^{(1)} = [t\,t\,t\,t\,t\,w\,w\,w\,w\,t]$ in Table 3.5. Let $S_j^{(i)}$ be the set of all *sub-sequences* of $MS_j^{(i)}$. Let sub_c be a MA subsequence that occurs at least once in the combination of all l instances of CA_i, i.e., $sub_c \subseteq \cup_{j=1,\ldots,l} S_j^{(i)}$.

Definition 3.4 (Cover of sub_c) Denoted as **cov**(sub_c, M_i), equals the number of instances in M_i that *contains* at least one instance of sub_c.

Definition 3.5 (Support of sub_c) Denoted as **supp**(sub_c, M_i), equals $\frac{cov(sub_c, M_i)}{l}$.

For example, column 3 in Table 3.5 shows the length-3 subsequence [t t w] occurs 2 times among 3 ($l=3$) instances of CA_1. Hence, cov([t t w],CA_1)=2 and supp([t t w],CA_1)=2/3.

Research work has revealed that patterns with low support in individual classes (each CA label is a class) or with very high support globally across classes are typically not useful for classification, as such patterns either occur very infrequently in the class instances or are a common occurrence in multiple classes, respectively [32]. In our scenario, an individual MA symbol (e.g., "sit") is likely to be very common in all instances, while a long sequence of MAs will have low *cover*.

There are many types of discriminatory patterns possible in such sequences (e.g., patterns with multiple wild cards) and defining new pattern mining algorithms is not our focus. Instead, we present two broad strategies that extracts *sub-sequences with high observed support.* These sub-sequences (sub_cs) may have high discriminatory power w.r.t. at least one other CA in the data.

Temporal Duration-preserving strategy (SA-TD). Given a minimum support threshold Θ_0 and a maximum sub_c size K_{max}, this strategy discovers the set of all sub_cs of length $[2, 3, \ldots, K_{max}]$, that have $supp(sub_c, CA_i) \geq \Theta_0$.

For example, column 3 in Table 3.5 shows that the sub_cs of length 3 selected with $\Theta_0 \geq 0.6$ for CA_1 are {t t w}, {t w w}, {w w t}. The SA-TD strategy finds the *union* of all such qualifying sub-sequences across all CAs in the training data. For example, in Table 3.5 this approach results in the selection of the following 3-element sub_cs as features: {t t w}, {t w w}, {w w t}, {t t s}, {t s s}, {s s t} across all the CAs (CA_1 and CA_2). Intuitively, SA-TD features capture a set of MA transitions among consecutive frames (including self-transitions, i.e., MA sequences of long duration) that are observed to occur often.

The resulting sequence features are appended to the OO features to create a longer $OO+Sequence$ feature vector, with the ith element of the vector corresponding to the frequency of occurrence of the corresponding feature. For example, for the instance $MS_1^{(1)}$ in Table 3.5, column 4 shows that the SA-TD approach results in a feature vector $[4, 0, 7, 1, 1, 1, 0, 0, 0]$, where the elements $[1, 1, 1, 0, 0, 0]$ come from the sub-sequence based features, as the sub_cs '[ttw]' & '[wwt]' & '[tww]' occur once each in $MS_1^{(1)}$.

Transition-preserving strategy (SA-TP). This approach preserves only the transitions between *distinct, adjacent MA*s, by removing (or collapsing) the run-length of consecutively repeating MA symbols for each CA instance. For example, column 5 in Table 3.5 shows this T-P sequence associated with $MS_1^{(1)}$ is transformed to [t w t]. By focusing purely on the sequence of *transitions* among distinct MAs, the SA-TP approach ignores slight variations in the duration of an individual MA and helps discover key underlying transitions. Consider an activity defined as a "smoking break." It is possible that two instances of this CA might share a certain latent sequence (e.g., defined by an order of $walk \rightarrow stand \rightarrow walk$), while differing slightly in the duration of each micro-activity (e.g., [w w w t t w] and [w w t t t w w]). The SA-TP approach would identify [w t w] as a potential unifying & discriminatory feature, and in addition, help to *reduce the dimensionality* of the resulting feature vector. With our data sets, SA-TP results in ~2–4 fold reduction in the size of the resulting feature vector, compared to SA-TD.

Feature Reduction. As the *SA* approaches have lead to high dimensionality of the feature vectors, the feature extraction step may be followed by a step of correlation-based selection of good features [145]. Subsequently, CA classification models may be built on the final reduced feature space, using labeled training data, to carry out prediction of unknown CA instances.

Regression Features

These features are based on the idea of building trend models to build discriminative features of complex activities and further make classification. The main intuition here is to learn the evolution of a certain CA in the underlying feature space K as the activity progresses in time. In our case, K

indicates the number of clusters or MAs. We use the features from the activity evolution models to build classifiers. For testing an unknown *CA* instance, the evolution model of the test stream is compared with known models to predict the *CA*. The intuition is explained in Figure 3.8.

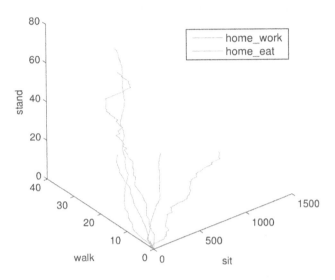

Figure 3.8: Regression features for activity learning.

In Figure 3.8, we observe the temporal evolution of five *CA* instances, with the feature space containing 3 locomotion labels (i.e., "sit," "stand," "walk"). Note the trend difference between two activity types, i.e., home_work and home_eat. Therefore, we can build the regression models of these *CA* instances, and use the regression coefficients to infer CAs. For the example of 3 locomotion labels ("sit," "stand," "walk"), we can build a linear regression model (based on three variables) for an activity class:

$$\beta_0 + \beta_1 a_1 + \beta_2 a_2 + \beta_3 a_3 = 0, \tag{3.3}$$

where a_1, a_2, a_3 indicate the number of "sit," "walk," "stand" frames in each *CA* instance. Thereafter, the coefficients $\langle \beta_0, \beta_1, \beta_2, \beta_3 \rangle$ can be used to train classification models.

Topic Features

A Topic model[4] is a well-known approach in text mining, for discovering the abstract "topics" for a collection of documents. This is based on analyzing the statistical properties of the "bag of words" in the document. For implementing topic models, Latent Dirichlet allocation (LDA) is one of the most well-known instantiations [15, 151].

Topic models have successfully been utilized in several applications including activity recognition [14, 80]. In activity recognition literature, researchers have typically applied topic models

[4]http://en.wikipedia.org/wiki/Topic_model

over multi-sensor streams, typically in constrained wearable computing setups, where rich observation data about the activity is available.

For our current classification task, we build latent topics from a single ACC stream, by transforming the stream to an observation space representing data clusters present in the stream. Thereafter, we would use the emerging topic associations in turn as features, to build supervised models for studying complex activity recognition accuracy.

In our problem, a topic t is a hidden latent activity. Elements in a given CA instance (i.e., the sequence $\langle l_1, \ldots, l_n \rangle$) are probabilistic occurrences due to $T (>= 1)$ hidden latent activities (act_{topic}). A complex activity CA is a statistical combination of a group of such latent activities or topics $\langle act_{topic_1}, \ldots, act_{topic_j} \rangle$, and each latent activity act_{topic} is associated with a certain probability. The intuition is, from the accelerometer observations, CAs such as *office_break* can consist of "getting off the chair," "moving around," and "sitting down on the chair," etc., *office_lunch* can include "carrying the tray to the desk," "sitting down," and "eating," etc. Our aim is to recover the resulting topic distributions, if any, as features to classify the CA.

To do this, we apply LDA to infer the latent topics (act_{topic}) from the CA sequences representing cluster labels (tag_{frame}). Then, based on the act_{topic} distributions for each complex activity ($act_{complex}$), the actual CA label for each instance is to be inferred. Therefore, we have the following inference for T latent topics:

$$p(act_{complex}|tag_{frame}) = \sum_{k=1}^{T} p(act_{topic}|tag_{frame}) p(act_{complex}|act_{topic}). \tag{3.4}$$

Once we have learned the LDA model, we can use this model to estimate the distribution of latent activities act_{topic} in a given CA instance, i.e., the estimate of extent to which different latent topics are present in the CA. Therefore, for each training CA instance, we have a vector $\langle p_1 \cdots p_T \rangle$, representing the distribution of the T latent topics in the CA instance. This vector represents the "*Topic Features*."

3.5.4 COMPLEX ACTIVITY LEARNING APPROACHES

Next, we describe two techniques to learn the CA models.

1-*Tier* Learning Approach

This approach extracts time and frequency domain statistical features directly from the *raw accelerometer data stream* associated with each CA instance. In this approach, for each training instance of CA_i, we consider the accelerometer readings recorded for that instance and compute the statistical orientation-dependent and independent features, just as we did for MA classification (Table 3.2). Thus, classifier models using supervised techniques are built from the training data, represented as a set of vectors containing statistical properties of the activity stream. CA labels are predicted using the model on unknown test data.

2-*Tier* **Learning Approach**

In this approach, we first convert the raw stream into a set of micro-activity (MA) labels. The MA label can be either the real micro-activity label like sit, stand, or the cluster label by unsupervised learning. Therefore, the complex activity recognition task is to detect a CA (e.g., cooking, taking-dinner, working) from the sequence of low-level MA labels. The complex activity label is noted as $l_i^{(CA)}$, corresponding to the accelerometer frame label $l^{(A)}$. We have $l_i^{(CA)} \in \mathcal{L}_{CA}$ as the set of CA labels.

Using the CA features described earlier, several machine learning models may be used now to learn the CA models from their MA-based representations. Here, we explore the support vector machine (SVM) method. In particular, we apply the LibSVM package which has a good performance in many classification applications [30]. Figure 3.9 captures the overall procedure for our complex activity learning and classification. The feature dimensions (*Set, Regression, Topic*) are essentially representing information about a *CA* instance in different dimensions. We describe two vector machine variants for learning classification models: *early* fusion and *late* fusion.

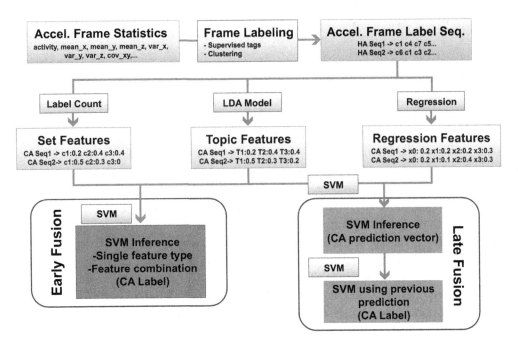

Figure 3.9: Learning procedure (early fusion vs. late fusion).

- **Early Fusion.** In early fusion, we directly build classifier (e.g., SVM model) on the *CA* features, and make prediction. Here, the *CA* features can be a single type of features (i.e., set, regression and topic) or a combination of these different types of features to form a single, master feature vector. From training data, these feature vectors are used to build a single SVM classification model. Intuitively, this approach learns the separation vector on the feature space, combining the multiple dimensions and can exploit correlations among features in the mixed feature space.

- **Late Fusion.** This approach exploits the discriminatory power of individual feature dimensions for classification, while compromising on the potential correlations present in the mixed feature space combining all feature extractors. Here, we learn *individual* SVM models for each feature dimension. Each SVM model outputs a prediction vector $[p_1, \ldots, p_n]$ for each *CA* instance, where p_i = predicted probability of the instance to belong to class i. Afterwards, the third vector plane is subsequently learnt using the prediction vectors of different *individual* SVMs, to predict the final class value of an unknown *CA* instance. Intuitively, this model accommodates variation in the predictive powers of multiple feature dimensions, and performs well when feature spaces are not necessarily correlated.

3.5.5 LEARNING PERFORMANCE

Now, let us discuss a few results of applying the above learning models on a real data set.

Learning approaches: 1-*Tier* **vs.** 2-*Tier*

Figure 3.10 presents the classification accuracy statistics observed in 1-*Tier* method and the four feature extraction strategies in the 2-*Tier* approach (i.e., OO, SA-TD, SA-TP and the combined feature sets SA-TD+TP) introduced earlier. We experiment with a variety of classifiers (*decision tree –J48, Adaptive Boost - Adaboost, LibSVM, Bayesian Network,* and *Naive Bayes*) and plot the mean, and the standard deviation of the measured accuracies across these classifiers. We observe that the 2-*Tier* approach results in an across-the-board improvement in the classification accuracy, ranging from 7- 20%, compared to the 1-*Tier* approach. Due to the different dynamics of lifestyle activities of different users, the absolute accuracy values are user-dependent. A salient observation is that even the OO approach, with a slim feature vector dimension (7 MAs), mostly out-performs the 1-*Tier* approach which uses about 70 statistical features (with correlation-based feature selection). We also note that sequence-based features provide an additional, but variable (4-15%) amount of improvement in the classification accuracy. These results establish the superior quality of locomotive signatures, compared to their statistical counterparts. Note that the 1-*Tier* statistical feature-based approach should provide almost comparable performance to 2-*Tier*, for those semantic activities that are dominated by a single locomotion. We observed this in some CAs that have dominating "sitting" states, like office work.

Figure 3.11 shows the performance of varying K_{max}, i.e., the maximum possible sequence length considered in SA-TD & SA-TP. The plot shows that the classification accuracy shows a marginal improvement initially, before flattening out at K_{max} beyond 3 or 4. This demonstrates that relatively short MA sequences possess the highest discriminatory power.

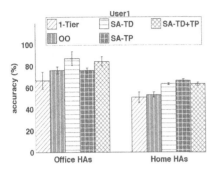

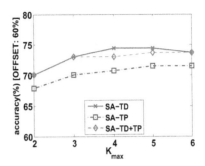

Figure 3.10: Performance 1-*Tier* vs. 2-*Tier*   **Figure 3.11:** Sensitivity analysis of K_{max}

Features Comparison: Set, Regression, Topic. Figure 3.12 provides the comparison among the three feature types. We report poor performance in recognizing both home and office activities using *Regression* features. Such regression features are computed using simple linear regression models fitted to the activity evolution. Between *Set* and *Topic* features, there does not seem to be a clear winner, though *Topic* features perform better in most cases (7 out of 10), with *User*2's home activities recording the highest gain ($\geq 25\%$) using *Topic* features. Overall, the average percentage improvement achieved using *Topic* features surpasses the *Set* features.

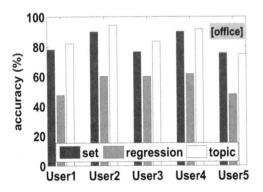

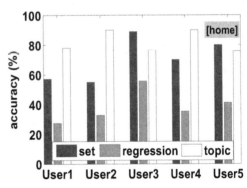

Figure 3.12: Performance comparison of *Set*, *Regression*, and *Topic* features. (10-clusters are used for this comparison.)

Learning: Early Fusion vs. Late Fusion. Finally, we evaluate and compare the two learning approaches, i.e., early fusion and late fusion, using the combination of multiple features. As *Regression* features cannot achieve good performance with the reality data, we mainly test fusion using the combination of *Set* features and *Topic* features. As shown in Figure 3.13, we observe that all home and office activities of five users can achieve good recognition accuracy using late fusion. The fact that late fusion works better indicates that perhaps not much correlation is present in the two feature dimensions (*Set*, *Topic*), for exploitation by the early fusion approach. The final accuracies computed by late fusion for home and office, are between 85–95%, as compared to ≈65–80% reported previously in Figure 3.10. The highest accuracy we achieved is 97%. The average accuracy stays around 86.17%.

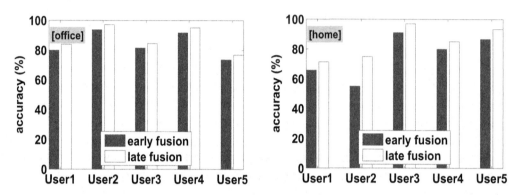

Figure 3.13: Comparison of early fusion and late fusion.

Table 3.14 provides the confusion matrices of both home and office activities of five users from the late fusion-based classification. We observe that office activities are relatively easier to detect compared to home activities. In particular, all *office_work* complex activities from 5 users have an impressive ≥90% accuracy. For home activities, we observe the classifier gets confused for *home_relax* activity for *User*1. In our exit interview with *User*1, we subjectively found that *User*1 often performed this activity in ways that intuitively should look similar to other activities like *home_cook* and *home_relax*, when observed solely by the accelerometer. Since the tagging process is unconstrained, it is quite possible that some activities are similar in the observation space of an accelerometer.

Overall, we find that the power of the ACC alone to discriminate between these CAs varies from user to user. We also observe that ACC adds a healthy discriminatory power towards classifying real-life CAs. Combined with other observations such as time-of-day or learning the activity sequences across days may boost these numbers further.

3.6 CONCLUSIONS AND SUMMARY

This chapter presented an overview of activity recognition using a single smartphone-resident accelerometer sensor. We discussed several factors that play a role, particularly in activity recognition in the wild, and with the smartphone accelerometer.

There are a few areas where research is concentrating its focus currently.

	Activity	No.	Confusion Matrix					Acc
USER1	home_work	36	**0.88**	0.0	0.04	0.0	0.08	71.4%
	home_break	14	0.0	0.4	0.2	0.4	0.0	
	home_relax	25	0.066	0.0	0.667	0.133	0.133	
	home_cook	21	0.0	0.0	0.2	**0.8**	0.0	
	home_eat	17	0.2	0.1	0.2	0.0	0.5	
	office_break	17	**0.9**	0.0	0.0	0.1		84.0%
	office_work	30	0.0	**0.96**	0.04	0.0		
	office_meet	15	0.1	0.8	0.1	0.0		
	office_lunch	11	0.0	0.0	0.0	**1.0**		
USER 2	home_relax	21	**1.0**	0.0	0.0	0.0		75.0%
	home_work	9	0.2	0.6	0.0	0.2		
	home_baby	9	0.4	0.2	0.4	0.0		
	home_eat	12	0.1	0.1	0.0	**0.8**		
	office_work	80	**0.98**	0.2	0.0	0.0		97.1%
	office_toilet	33	0.067	**0.866**	0.067	0.0		
	office_lunch	19	0.1	0.0	**0.9**	0.0		
	office_meet	20	0.4	0.6	0.5	**0.85**		
USER 3	home_relax	57	**0.98**	0.02	0.0	0.0		97.0%
	home_cook	23	0.05	**0.9**	0.05	0.0		
	home_eat	28	0.0	0.4	**0.96**	0.0		
	home_clean	11	0.0	0.2	0.0	**0.8**		
	office_work	122	**0.954**	0.09	0.181	0.0	0.181	84.5%
	office_lunch	32	0.167	0.7	0.0	0.033	0.1	
	office_coffee	35	0.2	0.04	0.56	0.16	0.04	
	office_toilet	33	0.067	0.0	0.066	**0.867**	0.0	
	office_break	15	0.2	0.0	0.6	0.0	0.2	
USER 4	home_relax	15	**0.813**	0.062	0.125			85.0%
	home_work	7	0.0	**1.0**	0.0			
	home_cook	18	0.1	0.0	**0.9**			
	office_meet	11	0.6	0.2	0.2			95.0%
	office_work	59	0.022	**0.978**	0.0			
	office_break	41	0.0	0.1	**0.9**			
USER 5	home_cook	20	**0.95**	0.0	0.5			93.3%
	home_work	6	0.333	0.667	0.0			
	home_relax	11	0.0	0.0	**1.0**			
	office_work	65	**0.927**	0.018	0.018	0.036		76.8%
	office_meet	11	0.0	0.667	0.333	0.0		
	office_lunch	23	0.733	0.0	0.067	0.2		
	office_break	45	0.16	0.0	0.04	**0.8**		

Figure 3.14: Confusion matrix of late fusion results.

1) **Online Learning & Early Detection.** Current approaches are largely limited to offline detection of activities with the complete sensor (e.g., accelerometer) data when the activity is ending. Online continuous detection of an activity is more challenging and useful in real-life applications. In order to detect an activity from a *streaming* data, one has to consider how the classification algorithms work in the presence of partial data. There is also work necessary to

understand how the trained models can be evolved as the activity pattern itself evolves with time.

2) **Concurrent and Interleaved Activities.** In a real-life setting, the end of one activity might be overlapping with the beginning of the next. Activities might also overlap completely. For example, if an user is having a coffee break from work at home, while also filling up clothes for laundry, the user is performing two activities in parallel. Some works [66] have started investigating these special cases, however, there is significant work necessary to bring such algorithms to perform in real-time, using observations from a smartphone platform.

3) **Cross-user and Large-scale Crowd Models.** In this chapter, the learned models either for micro-activity or for complex activities are user specific. There is a body of research that is focusing on large-scale models that may be used across multiple users. It is non-trivial to build such models due to the data diversity. Some initial works [1] have started looking into this aspect.

CHAPTER 4

Energy-Efficient Computation of Semantics from Sensors

Smartphones require a significant amount of energy due to their increased processing power, embedded sensors, network connectivities, and a large number of power hungry applications. Due to the restricted phone size and limitation of battery technologies, energy consumption becomes a fundamental bottleneck for smartphone daily usage. In the previous chapters, we presented a set of key techniques for computing semantics using phone based mobile sensing, like building semantic trajectories from GPS, extracting semantic activities using accelerometer data, etc.

Computing such semantics using data from mobile sensors requires continuous sampling of sensor data. Continuous sampling leads to high battery drain and will significantly hamper user experience. Therefore, we need mechanisms to reduce the sampling frequency, and strike a balance between energy requirements and application's requirements to derive the semantics. Sustainability and energy conservation are important topics for economical, practical, and environmental reasons. In this chapter, we present key techniques for building energy-efficient methodologies for computing semantics using the smartphone.

4.1 ENERGY-EFFICIENT MOBILE SENSING

In order to better understand a smartphone's energy limitations, we first review the specification of energy consumption of mobile sensors. In particular, we present a detailed overview of the sensing cost for the various sensors available on a smartphone. Next, we discuss methods to build energy-efficient sensing methodologies, from both hardware and software perspectives.

4.1.1 SMARTPHONE BATTERY LIMITATIONS

Due to the rapid progress of smartphone technology, there has been significant amount of interest in profiling the energy consumption of a smartphone [26, 44, 52, 129]. The main objective of these studies was to better understand battery limitations and gain insights on energy consumption on a smartphone.

Perrucci et al. [129] provided a comprehensive survey on the distribution of phone energy consumption. They divided the energy consumption into three main categories: wireless communication (e.g., WiFi, Bluetooth, 2G/3G networks), mobile services (e.g., voice, video, SMS), and miscellaneous (e.g., CPU, display, mobile TV, memory). This study is conducted on the Nokia

N95 phones using the built-in power profiling. They discovered that the most energy-hungry part of a mobile phone is wireless communication, as opposed to high-definition displays or multi-core CPUs. Furthermore, the experiments showed that the top six energy consumers on a Nokia N95 phone are (in descending order): downloading data using a 3G network, downloading data using WiFi, sending an SMS, making a voice call, playing an MP3 file, and displaying backlight. In general, we could say that networking typically requires more energy compared to computation and processing. In [44] the authors present an evaluation of the energy consumption by different typical sensors on the smartphone. This study presents statistics on power consumed by these sensors in terms of hours and the battery usage percentage (see Figure 4.1 for details).

Test	Power [mW]	Time (Energy) to get position [s] ([J])				Time needed to shutdown [s]	Precision [m]	Battery life [hour]
last call		< 10s	1min	10min	1h			
		Positioning methods						
1. GPS	~356	3 (1.07)	30 (10.7)	40 (14.2)	200 (71.2)	30	10	11.5
2. AGPS 3G	~973	11 (10.7)				30	10	11.5
3. AGPS 2G	~765	12 (9.18)				30	10	11.5
4. Net 2G	~663	3 (1.99)	6 (3.98)	6 (3.98)	6 (3.98)	6	100 - 1000	6.2
5. Net 3G	~1150	3 (3.45)	5 (5.75)	5 (5.75)	5 (5.75)	3	100 - 1000	3.7
		Sensors						
6. Accelerometer	~124	NA				0	NA	28.1
7. WLAN	~800	0.85 (0.68)				0	100	38.7
8. Cell-ID	~0	0				0	1000	58 - 75
		Standby						
9. Online 2G	~60							74.8
10 .Online 3G	~77							58.3
11. Offline	~36							124.7

Figure 4.1: Smartphone energy consumption specification [44].

Similarly, Carroll [26] provided specific analysis of the power consumption of a smartphone's hardware components, including CPU, memory, cellular, GPS, graphics, LCD, SD card, Bluetooth, Wifi, etc. Together with previous studies, it shows that GPS and WiFi consume a relatively higher power, compared to other short-range communication protocols like Bluetooth. Recently, the latest Bluetooth LE (low energy) technology has gained attention in many areas such as healthcare, fitness, security, and home entertainment. One of the reasons for the popularity is its energy efficient operation.

Next, we give an short overview of the hardware approaches behind energy-efficient mobile sensing.

4.1.2   ENERGY-EFFICIENT SENSING: HARDWARE APPROACHES

There has been significant research contributions in building energy-efficient system architectures to preserve the battery life of smartphones while it is performing sensing activities. [157] highlights the fact that it is beneficial for various sensor units along with their own memory and processors perform their functions independently (e.g., turn on/off instantly without waiting for other units) in order to save energy.

From the design perspective, a co-processor sitting on a sensor unit to process the (usually) analog data coming from the sensor, has three power states: *run*, *idle*, and *sleep*. The *run* state consumes energy as a function of the frequency of the processor—higher the processing frequency, higher is the power required. It is also the highest energy consumer compared to the *sleep* and *idle* states. The *sleep* state consumes less energy compared to the *idle* state. As a consequence, research works have tried to maximize the amount of time the co-processor spends in the *sleep* state. For example, the authors in [132] design an energy-efficient continuous sensing architecture called *LittleRock*. LittleRock decreases the energy overhead of continuous sensing by using a dedicated low-power co-processor that performs energy efficient sampling and low-level processing of sensor data.

Additional details on energy-efficient sensing can be found in works like [104, 132] and others mentioned above, and are beyond the scope of this book. In the following sections, we will focus on presenting software-based energy conservation methods for mobile sensing.

4.1.3   ENERGY-EFFICIENT SENSING: SOFTWARE APPROACHES

In order to minimize the energy drain, the software-based approaches focus on building algorithms to reduce the duty cycle and/or sampling frequency[1] while performing continuous sensing. Here, we provide a brief overview of different types of energy-efficient sensing approaches from the software perspective. We perform a couple of case studies to describe details of two select techniques along the two broad dimensions we focus on in this chapter.

At a high level, apart from the personal use of smartphones, these devices are also envisioned to act as a channel for affixing additional sensors such as air quality or humidity, temperature, and for dedicated work force personnel to use it for different application scenarios. Smartphones along with its in-built sensors and such associated accessory sensors are able to continuously monitor an environment or a phenomenon. In its current working model, the community (users) act as the carrier of these sensors. This phenomenon of using the community to sense is referred to as *community-driven sensing* or *community sensing* or *participatory sensing*. *Participatory sensing* puts more focus on modeling the user's participation semantics (When does the a user want to participate? How? Etc.). At its core, community sensing is a dynamic new form of mobile geo-sensor network [2].

Along these lines, one body of research investigates how effective and efficient such a "swarm of smartphones" may be, to sense a general phenomenon of interest. Examples of such

---

[1]http://en.wikipedia.org/wiki/Duty_cycle

phenomenon could be noise pollution, air quality or traffic estimates of a city, or pollen/humidity spreads in a vast agricultural field. A key problem over here is: *How do I place the sensors and move the users so that I can maximize my coverage while minimizing the number of sensors and the total energy consumed?* This problem has been investigated in traditional wireless sensor network research community. The general direction is to devise sensing strategies (where to sense? when?) to maximize the coverage and minimize the number of sensors required. This is known as the *optimal sensor placement* problem. As such, the problem of placing K sensors optimally in an arbitrary sensor field and finding the optimal sampling frequency is hard to solve. Some of the sub-problems are either NP-complete or NP-hard [38, 92, 101, 102], and thus a number of near-optimal algorithms have been proposed in these studies.

For example, [92] provides an excellent overview of the theory behind optimal sensor placement policies, and discusses near-optimal/greedy solutions using Gaussian processes. In [101], the sensor placement is modeled as a min-max optimization problem that is solved using an algorithm based on simulated annealing.[2] Similarly, [38] proposes a sensor detection strategy while considering aspects like terrain properties. [102] proposes a method for sampling regions of interest, and perform information aggregation from the multiple sensors locally, to build an adaptive sampling strategy.

As a natural extension, application of these works in traditional sensor networks have been investigated for community sensing [27, 165]. However, unlike traditional sensor networks, in community sensing the sensors are users, performing their own operations, and do not necessarily follow a sensing pattern that would improve coverage or reduce redundancy. Hence, this environment is comprised of un-controlled or semi-controlled mobility of sensors. Sensor readings might not always be reliable. [2] provides a good overview of the various challenges associated with such a form of community-driven sensing. However, the fundamental problem has not changed. Most of the works trying to address this space focuses on how to reduce sampling frequency, while not compromising on coverage and quality of the models of the phenomenon being built.

Recently, there is also work starting on how we can combine personal context to determine: *Is it the right time for me to sense a phenomenon?* For example, air quality or sound pollution can be effectively sensed when the smartphones are on the road, as opposed to in the car. Along these lines, works [121] that focus on efficient extraction of personal context are important to effectively use the community to improve sensing quality of a phenomenon.

Solutions along the lines of exploiting the community as a whole build collaboration mechanisms between various smartphones or different sensors within the phone effectively sense a phenomenon. For example, authors in [158] investigate sensing using multiple autonomous mobile devices, in which the natural mobility of these devices can be turned into an advantage over static sensing as they can cover larger areas and reduce energy consumption. Similarly, the OpenSense project [2] applies this principle by mounting air quality monitoring sensors on buses and trams. A global scheduling and placement problem is defined by Saukh et al. [143]. Their solution is to

---

[2]http://en.wikipedia.org/wiki/Simulated_annealing

select the tram lines and the sampling points that provide the best coverage of a city. Human subjects on the other hand typically has unconstrained mobility [76] and keep performing personal activities. These need to be considered while designing scheduling algorithms when smartphones of users are the "community sensors."

A body of work studies the tradeoff between *what is sampled* and *at what cost*. For example, the Kobe toolkit [34] focuses on balancing the accuracy and energy cost of activity classification algorithms on mobile devices, by dynamically adjusting the pipeline of sampling frequency, feature extraction and machine learning components. The runtime adaptation focuses on adjusting where the computation is executed (device vs. cloud) and takes into consideration system changes, like changes in phone battery levels or foreground processing load. The SociableSense system [136] is another example of a work in this area that applies reinforcement learning techniques to adjust the duty cycle of multiple sensors and uses a multi-criteria decision theory. It transfers a few computational tasks from the device to a back-end cloud.

For the rest of the chapter, we restrict ourselves to describing a couple of case studies from our prior work, on software approaches towards energy conservation. These works are along two major dimensions: (1) How to sense a phenomenon like air quality using vehicle-mounted sensors as the "community" of sensors? (2) How to sense personal activity routines? Both the studies focus on answering these questions while achieving better energy efficiency than typical algorithms.

4.2   MODEL-BASED ENERGY-EFFICIENT SENSING

Consider the case where we have air quality sensors mounted on community-driven vehicles like buses and cars. Environmental sensors are deployed on public transport buses to monitor air quality (using sensors to measure CO_2, CO, NO_2, etc.). Our main objective is to come up with strategies of how to best sense the phenomenon, while keeping the energy budget under limit. Let us call this "optimal sensor placement," although in the strict sense of the word "optimal," the problem is NP-complete (as described earlier). Hence, our solution also falls around the classes of solutions that use near-optimal or greedy strategies. We have two levels of objectives: (1) build an optimal sensing strategy for an individual bus line, i.e., determining when and where each bus should take a measurement and (2) build a globally optimal sensing across all bus lines, such that the total energy consumed due to sensing activities is under a limit. Figure 4.2 shows a sample sensing strategy from four bus lines, where the symbol "×" indicates the points where the buses record sensor samples. If the buses collaborate with each other, then we can reduce the sampling frequency of certain sensors. For example, *Bus* 3 can have only two sensing points (blue "×"), as there are overlapping routes with other buses, who have already taken measurements on the overlapped routes.

In the above scenario, our objective is to maximize the sensor coverage while keeping the energy cost due sampling below a threshold (i.e., ideally "minimize" the cost). This problem is challenging compared to prior sensor network set-ups where sensors were static, or even if they

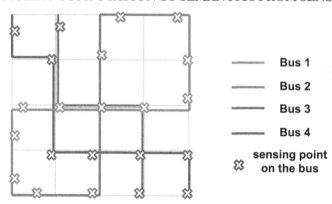

Figure 4.2: Air quality sensors mounted on moving buses can collaborate with each other to reduce energy costs.

were mobile, top-down policies were used to guide where sensors should move. In our case, the community of sensors is autonomous and mobile.

For this situation, we seek to design an optimal mobile sensing strategy, which offers a tradeoff between coverage maximization and sampling cost minimization of the community of mobile sensors. This raises two key questions.

a) For a short sensing duration (e.g., a bus moving for 2 h), how can you find an optimal sensing (or sampling) protocol[3] that can guarantee required sensor coverage with a limited sensing cost budget?

b) For a long sequence of sensing durations (e.g., a bus moving for one month or several days), how can you find an optimal sensing protocol that samples only when necessary? Does segmenting the trajectory help us devise a better mobile sensing strategy?

To address these questions, we present a model-driven sensing strategy. This is particularly suited for community-driven mobile sensors. We call it *OptiMoS*. It is a two–tier mobile sensing framework that uses learning techniques like linear regression, support vector regression, etc. In the first tier, a long sensing sequence (e.g., the complete route of a bus line) is divided into several non-overlapping segments, where the data points in each segment are "homogeneous" (discussed later). In the second tier, OptiMoS chooses optimal sampling points for each segment, as these data points can provide maximum sensor coverage.

For segmenting sensor data from mobile sensors, we design and compare different segmentation algorithms. These include the exhaustive search, dynamic programming, binary top-down segmentation, error-tolerated heuristic segmentation, etc. For specifying the sampling frequen-

---

[3]"protocol" and "policy" are used synonymously here

cies in each segment, we design and compare several sampling algorithms: uniform, random, error-based entropy sampling, and mutual information sampling.

4.2.1 TWO-TIER OPTIMAL SENSING

This optimal sensing problem can be formulated as follows. *Given a sequence of N mobile sensor readings* $\mathcal{R} = \{R_1, \cdots, R_N\}$ *from continuously moving sensors, where each record* $R_i = \langle t, l, x_1, \cdots, x_M \rangle$ *consists of M types of sensor readings (from* x_1 *to* x_M*) together with the timestamp (t) and the location (l = ⟨longitude, latitude, altitude⟩), the objective of OptiMoS is to identify the best sampling of such sequence of sensor readings that can guarantee the majority of sensor reading information (i.e., sensor coverage maximization) at the minimum sampling rate (i.e., energy cost minimization).*

Figure 4.3 sketches the framework and system architecture of OptiMoS.

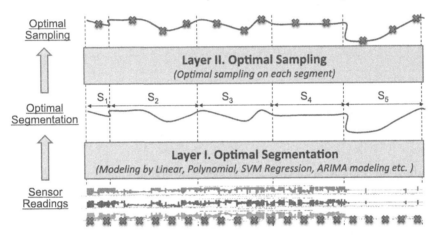

Figure 4.3: OptiMoS's two-tier optimal sensing framework.

The lower tier of OptiMoS consists of the initial input in the form of raw sensor readings collected by the moving sensors. This is in the form of multi-dimensional spatio-temporal data. Each record R_i, denoted by the "×" symbol in Figure 4.3, includes sensing time t, sensing location l (typically ⟨*longitude, latitude*⟩ from GPS), and environmental measurements x_1 to x_m.[4] The objective of this tier is to find the optimal segmentation (detailed later) based on data model on these raw readings. OptiMoS supports several kinds of modeling methods, e.g., simple linear regression, polynomial regression, SVM (Support Vector Machine) based regression, time series ARIMA (Auto-Regressive and Moving Average) modeling. From the first tier, we can achieve an optimal (or near optimal) segmentation (e.g., the five segments S_1 to S_5 in Figure 4.3).

The upper tier of OptiMoS focuses on studying segments that are computed from the lower tier. For each segment, the objective here is to select only a subset of sensor readings ("×" symbols

---

[4]For example, x_1 is the measurement of CO_2 concentration, x_2 is of CO, x_3 is NO_2, etc.

in Figure 4.3 in the top layer). This subset can keep enough modeling information for regression of the whole segment and for prediction of non-selected sensor readings. Consider Figure 4.3, from segment S_1 to S_5 we respectively keep only 1, 3, 3, 2, 3 numbers of readings to achieve reasonable sensing quality. For this optimal sensing example, OptiMoS only requires 12 sensing points instead of the initial 21 points. Thus the reduced sampling rate is $1 - \frac{12}{21} = 43\%$.

Tier I: Optimal Segmentation

The first tier of OptiMoS defines and determines the optimal segments. It defines optimal segments as the best K segments (i.e., $\mathcal{R}_1, \mathcal{R}_2, \cdots, \mathcal{R}_K$) such that the sum of the model errors for individual segment is minimized. A model $\mathcal{M}$ on a segment $\mathcal{R}_i$ can be linear, polynomial, SVM regression, ARIMA, etc. We empirically study linear and SVM regression, and evaluate their performances; other models could have similar principle. We apply the RSS (*Residual Sum of Squares*) to quantify the error for modeling a sequence $\mathcal{R} = \{R_1, R_2, \cdots, R_N\}$ (see Equation 4.1). The residual $res(R_i)$ of a reading R_i is the difference between its real value R_i and the approximation $\hat{R}_i$ obtained using the model $\mathcal{M}(\mathcal{R})$:

$$RSS(\mathcal{M}(\mathcal{R})) = \sum_{i=1}^{N}(res(R_i))^2 = \sum_{i=1}^{N}(|R_i - \hat{R}_i|)^2$$
$$\text{where } \hat{R}_i = \mathcal{M}(\mathcal{R})|_{R_i}. \tag{4.1}$$

Finding the best K segments is equivalent to identifying the best K-1 division points $R_{d_1}, R_{d_2}, \cdots, R_{d_{K-1}}$. Then, for each segment $\mathcal{R}_i$, we have a sub sequence of readings between two division points, i.e., $\mathcal{R}_i = \{R_{d_{i-1}}, R_{d_{i-1}+1}, \cdots, R_{d_i}\}$. For the first segment $\mathcal{R}_1$, R_{d_0} denotes the first reading R_1. Now, this optimal segmentation problem becomes an unconstrained optimization problem as follows:

$$\underset{d_1, d_2, \cdots, d_{K-1}}{\text{argmin}} \sum_{i=1}^{K} RSS(\mathcal{M}(\{R_{d_{i-1}}, \cdots, R_{k_{d_i}}\})). \tag{4.2}$$

The segment number (K) is not known in advance and should be estimated as a part of the optimization problem, as shown in Equation 4.3:

$$\underset{K, d_1, d_2, \cdots, d_{K-1}}{\text{argmin}} \sum_{i=1}^{K} RSS(\mathcal{M}(\{R_{d_{i-1}}, \cdots, R_{k_{d_i}}\})). \tag{4.3}$$

For simplicity, in the first step of this work, we can assume K is given and we will test a reasonably small set of different K values (e.g., $K \leq 10$), to analyze the convergence of segmentation algorithms and test an optimal segmentation.

Tier II: Optimal Sampling

After obtaining the optimal segmentation, OptiMoS identifies the best sampling for each individual segment. To quantify the quality of a sampling sequence $\mathcal{R}_{sub}$, we define *information loss* $\mathcal{L}(\mathcal{R}, \mathcal{R}_{sub})$ that captures the *RSS* increase between $\mathcal{R}_{sub}$ and the complete set of readings $\mathcal{R}$ as follows:

$$
\begin{aligned}
\mathcal{L}(\mathcal{R}, \mathcal{R}_{sub}) &= \frac{RSS(\mathcal{M}(\mathcal{R}_{sub} \to \mathcal{R})) - RSS(\mathcal{M}(\mathcal{R}))}{RSS(\mathcal{M}(\mathcal{R}))} \times 100(\%) \\
&= \frac{\sum\limits_{i=1}^{N}(R_k - \mathcal{M}(\mathcal{R}_{sub})|_{R_k})^2 - \sum\limits_{i=1}^{N}(R_k - \mathcal{M}(\mathcal{R})|_{R_k})^2}{\sum\limits_{i=1}^{N}(R_k - \mathcal{M}(\mathcal{R})|_{R_k})^2},
\end{aligned}
\tag{4.4}
$$

where $RSS(\mathcal{M}(\mathcal{R}_{sub} \to \mathcal{R}))$ is the RSS error for the approximation of the complete sequence $\mathcal{R}$ using the model $\mathcal{M}(\mathcal{R}_{sub})$ that was estimated from the sub sequence $\mathcal{R}_{sub}$.

Similar to the sensor placement problem in static WSN, there are two ways to represent the optimal sampling problem in OptiMoS: (1) Given a limited sampling rate δ, find the best sampling set $\mathcal{R}_{sub}$ that has minimum information loss $\mathcal{L}(\mathcal{R}, \mathcal{R}_{sub})$; and (2) Given an acceptable information loss threshold ϵ between sampling sub-sequence $\mathcal{R}_{sub}$ and the complete sequence $\mathcal{R}$, find the best sampling points such that the sampling rate is minimized. These two problems can be formulated as the following optimization problems:

$$
\underset{\mathcal{R}_{sub}}{\text{argmin}} \quad \mathcal{L}(\mathcal{R}, \mathcal{R}_{sub}) \quad s.t. \quad |\mathcal{R}_{sub}|/|\mathcal{R}| \leq \delta \tag{4.5}
$$

$$
\underset{\mathcal{R}_{sub}}{\text{argmin}} \quad |\mathcal{R}_{sub}|/|\mathcal{R}| \quad s.t. \quad \mathcal{L}(\mathcal{R}, \mathcal{R}_{sub}) \geq \epsilon. \tag{4.6}
$$

The sampling rate $|\mathcal{R}_{sub}|/|\mathcal{R}|$ denotes the cost of mobile sensing. Thus, we need to balance the information loss caused by the lack of sensor coverage with the sampling frequency (i.e., energy cost). The constrained optimization problems in Equations 4.5 and 4.6 can be rewritten as an unconstrainted optimization in Equation 4.7, by using a balancing coefficient λ as follows:

$$
\underset{\mathcal{R}_{sub}}{\text{argmin}} \, \mathcal{L}(\mathcal{R}, \mathcal{R}_{sub}) + \lambda |\mathcal{R}_{sub}|. \tag{4.7}
$$

4.2.2   MODEL-BASED OPTIMAL SEGMENTATION

This section presents various optimal segmentation strategies for solving the optimization problem presented in Equation 4.2. The strategies include search using dynamic programming, top-down binary segmentation, error-based heuristics, and near-optimal segmentation.

Optimal Segmentation

The segmentation problem in Equation 4.2 can be solved using *dynamic programming* (DP) [11, 77] with the complexity of $O(KN^2)$ where K is the segment number and N is the number of points in $\mathcal{R}$. DP's quadratic complexity makes it impractical for segmenting real-life large scale mobile sensing data. However, on the other hand, note that we can apply DP to segment a short sequence with $K \leq 5$, and evaluate other segmentation methods using the optimal modeling error from produced DP. The DP algorithm for segmenting the mobile sensor readings is summarized in Algorithm 4.1.

We can use $segmentDP(\mathcal{R}, 1, N, K)$ to solve the problem defined in Equation 4.2. To speed up the execution, we provide an extra condition on "bestRSSBound" (recording the best RSS error at each recursive step) to filter out some unnecessary invocations $segmentDP$. It is

---

Algorithm 4.1: segmentDP $(\mathcal{R}, i, j, k)$

---

    input : $\mathcal{R} = \{R_1, R_2, \cdots, R_N\}$ // mobile sensor readings

              i, j // to make next segmentation in $\langle R_i, R_j \rangle$

              k // the number of segments

    output: *optimalRSS* // model error by optimal segmentation

1 /* find the optimal segment */

2 **if** $k = 1$ *and* $j > i$ **then**

3   print i, j; // print optimal sub-segments

4   **return** RSS$(\mathcal{R}, i, j)$; // compute model error RSS $\langle R_i, R_j \rangle$

5 /* impossible to find the optimal segment */

6 **if** $j - i < k$ **then**

7   **return** ∞;

8 /* recursive segmentation (from k to k-1) */

9 *optimalRSS* $\leftarrow \infty$; // initial the optimal RSS found so far

10 **foreach** $id \in [i + 1, j - 1]$ **do**

11   *firstSegRSS* $\leftarrow$ RSS$(\mathcal{R}, i, id)$;

12   *restSegRSS* $\leftarrow$ segmentDP $(\mathcal{R}, id, j, k - 1)$;

13   *totalRSS* $\leftarrow$ *firstSegRSS* + *restSegRSS*;

14   *optimalRSS* $\leftarrow$ min$\{$*optimalRSS*, *totalRSS*$\}$;

15 **return** *optimalRSS*;

---

worth noting that our objective is to find a segmentation that is not only optimal for the training data, but also applicable to new sequences of mobile sensing data. Consider the case of the OpenSense project[2] where sensors on the top of buses are continuously providing a time series of air quality readings. Here, the segmentation results on one day of *Bus-line*-1 should be generally consistent with the time series reported from other days on the same line. However, due to diurnal and spatio-temporal variations in the air quality, it might be difficult to achieve. Generally, we should experimentally determine how much training data we need for each city because different cities have different variations in the air quality readings.

Top-Down Binary Segmentation

Since the DP-based approach is impractical for long sequences of sensor readings, that are common in real-life, greedy top-down segmentation methods like the binary split method [86] have been proposed in the literature. The idea is to hierarchically split a sequence with maximum error into two sub-sequences, until the number of segments reaches K. We denote the top-down binary segmentation algorithm as *Binary*.

Binary always chooses a segment with the maximum model-based regression error (*RSS*) to perform further segmentation. Obviously the segments with more *RSS* will be divided into more sub-segments than segments with less *RSS*. As a result, the segmentation approach could lead to imbalanced segments. This situation is similar to worst case binary trees, where a tree is highly un-balanced and resembles a linked list. To overcome this situation we design an extended algorithm of *Binary* denoted as *Binary*$^+$.

Binary$^+$ proposes an *RSS* error measurement that introduces two types of penalties to prevent *Binary* from always choosing the segment with maximum *RSS* error: (1) How much error is reduced after splitting a segment into two? (2) What is the length of the resulting new segments (denoted as *length*)? The first penalty is to induce evaluation of segments that might benefit more with a splitting than one having high *RSS* error. The second penalty is to avoid splitting short segments even further. The *Binary*$^+$ algorithm is presented in Algorithm 4.2. As shown in Equation 4.8, α is the penalty coefficient.

$$newRSS = RSS(\mathcal{M}(\mathcal{R}_{left})) + RSS(\mathcal{M}(\mathcal{R}_{right}))$$
$$\widehat{RSS} = RSS(\mathcal{M}(\mathcal{R})) - newRSS + \alpha \times length. \tag{4.8}$$

Heuristic Segmentation

The *Binary* and *Binary*$^+$ algorithms chose a segment for division using *RSS* or $\widehat{RSS}$, but the point at which a segment is divided is always chosen to be the middle point of a segment. This might not be the best split. Therefore, we additionally design error-based greedy methods that use the model residual of each record to identify segment division. Equation 4.1 has described previously how to compute this residual. We propose heuristic methods that use a greedy strategy of using the point having maximum residual error as the division point for dividing the segments. Then, we recompute the new models for new segments recursively and find the next top error point as the new division point until we reach K segments. We denote this segmentation policy as *Heuristic*.

Similar to the *Binary*$^+$ algorithm, it is possible to design an extended version of *Heuristic*$^+$ called *Heuristic*$^+$ that uses the penalty function in Equation 4.8 to avoid forming short segments.

Algorithm 4.2: segmentBinary$^+$ $(\mathcal{R}, K)$

---

    input : $\mathcal{R} = \{R_1, R_2, \cdots, R_N\}$ // mobile sensor readings
           K // the number of segments
    output: *segOrderQueue* // list of segments

1 /* initial the segment result */
2 *segOrderQueue* $\leftarrow$ $\varnothing$;
3 /* insert the first segment to the sorted queue */
4 *segOrderQueue*.insert($\mathcal{R}$, 1, N);
5 **foreach** $id \in [2, K]$ **do**
6   /* retrieve & remove top error segment from the queue */
7   *topErrorSeg* $\leftarrow$ *segOrderQueue*.poll();
8   /* divide the segment into two subsegments */
9      $S_1 \leftarrow (\mathcal{R}, topErrorSeg.begin, topErrorSeg.division)$;
10     $S_2 \leftarrow (\mathcal{R}, topErrorSeg.division, topErrorSeg.end)$;
11   /* add the two new sub segments into two the sorted list*/
12   calculate $\widehat{RSS}$ for S_1 and S_2;
13   *segOrderQueue*.insert(S_1);
14   *segOrderQueue*.insert(S_2);
15   *segOrderQueue*.resort(); // resort for next segmentation
16 **return** *segList*;

In addition, *Heuristic*$^+$ does not search for the point having largest residual error, but for the longest contiguous sequence of points where the errors in each point exceeds a certain threshold (e.g., the error median). It then randomly chooses one of its end points of this segment.

The last segmentation approach we propose in OptiMoS is B^+H^+. This approach combines *Binary*$^+$ and *Heuristic*$^+$. The idea is to consider both the maximum error segment to build segmentation and the maximum error point for division. This combination is achieved by choosing the candidate segment to divide using *Binary*$^+$ and then within the chosen segment *Heuristic*$^+$ is applied to find the segmentation point. Since the segmentation points are in a region where the current model has its worst performance, we expect that the resulting segments have a uniform distribution. We compare the proposed segmentation algorithms using a metric called the *RSS_Ratio*. *RSS_Ratio* captures the error reduction obtained after segmentation and is defined as:

$$RSS\_Ratio = \frac{\sum_{i=1}^{K}(RSS(\mathcal{R}_i))}{RSS(\mathcal{R})} \times 100(\%). \tag{4.9}$$

4.2.3 MODEL-BASED OPTIMAL SENSOR SAMPLING

In this section, we discuss the second tier of OptiMoS. This tier deals with deciding on how to sample data from each segment created in the first tier. Below, we describe a few generally greedy strategies for performing this task.

Let (N) = total number of sampling points possible in a segment. Let δ = sampling rate selected. This means that with a sampling rate of δ, δN points are selected. Next, we discuss the different methods for selecting the sampling points.

Distribution-Based Sampling

The sampling policies can use statistical distributions (e.g., *uniform* or *normal*) for deciding on the sampling points.

- **Uniform sampling.** Here δN is selected uniformly from the segment. The gap (time or number of points) between two consecutive sampling points is *uniform* or *normally* distributed.

- **Random Sampling.** In this method we randomly select δN points from a segment.

Observe that the above policies considers only the distributions of positions. Each sampling point has an equal chance of being selected; thus, the sampling policy is unbiased.

Entropy-Based Heuristic Sampling

This method uses the residual $res(R_i)$ of each point in two different ways as the entropy for the selection of sampling points. Let us recall that every point has a residual value given by Equation 4.10. In the first method, we simply select the absolute residual and a threshold to select a set of sampling points. For example, the segment shown in Figure 4.4 initially has 25 sensor readings indicated by the blue times symbol ("×") at the bottom. The black line on the plot shows the residuals for each sensor reading. By using these residuals, we can pick top seven sensors (indicated by the red circles ("◯")), where three of them are primarily at the beginning of the segment and the four others are at the end.

In the second method, we use the *relative* residual to decide whether to select or not select a sampling point. To determine the relative residual $r\hat{e}s(R_i)$ of point R_i, first the residuals of its adjacent points are used to interpolate a value for R_i. Adjacent points are selected based on a window of size w around R_i. Let us call this interpolated value $leaveOneOutAppr(R_i, w)$. Thereafter, the relative residual $r\hat{e}s(R_i)$ is the difference between the actual residual $res(R_i)$ and the interpolated value. Different interpolation functions may be used to perform this interpolation (e.g., using basic mean, linear, Gaussian).

$$r\hat{e}s(R_i) = res(R_i) - leaveOneOutAppr(R_i, w) \qquad (4.10)$$

Figure 4.4 shows a visual explanation for a window size $w = 4$. After the relative residuals are computed, seven sensor readings are selected, shown using red rectangles ("□").

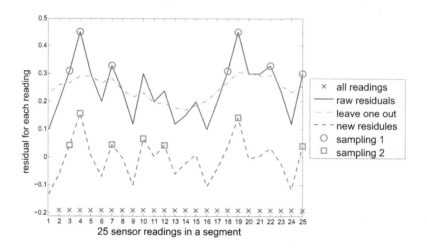

Figure 4.4: Entropy-based sensor readings sampling.

Mutual Information-Based Heuristic Sampling

In entropy-based heuristic sampling, both the absolute error *res* and the relative error *rês* are computed in advance and are not recomputed during the sampling process. However, the entropy of sampling points varies between each sampling step. We model this change using the *mutual information*, which measures the mutual dependence of two random variables. In our case it measures the dependence between the selected and the unselected points. Mutual information can reduce information dependency, therefore it has been largely used in many topics like feature selection [127] and static sensor placement [92].

Different from the earlier entropy-based sampling methods that directly select points with highest relative residuals (*rês* in Equation 4.10), a recursive method is used for computing $\widetilde{res}$ of the sensor readings. Our method removes the mutual information from sensor readings already selected from $\widetilde{res}$ as follows:

$$\widetilde{res}(R_i; \mathcal{R}_{sub}) = \hat{res}(R_i | \mathcal{R}_{sub}) - \hat{res}(R_i), \tag{4.11}$$

where $\mathcal{R}_{sub}$ is a set of sampling points already selected, R_i is a sensor reading that has to be added to $\mathcal{R}_{sub}$, $\hat{res}(R_i | \mathcal{R}_{sub})$ is the relative residual computed using the selected readings $\mathcal{R}_{sub}$, and $\hat{res}(R_i)$ is the relative residual computed using all the sensor readings $\mathcal{R}$.

The algorithm that uses mutual information for near-optimal sampling is presented in Algorithm 4.3. We start by applying the relative error-based entropy to select the first sampling points. Thereafter, we recursively compute the mutual information between candidate sample points and the selected sample points and use recursion to keep reducing the sampling points till the number of points reaches δN.

Algorithm 4.3: samplingMutualInfo $(\mathcal{R}, \delta)$

---

input : $\mathcal{R} = \{R_1, R_2, \cdots, R_N\}$ // mobile sensor readings
$\quad\quad\quad \delta \quad$ // the percentage of sampling
output: $\mathcal{R}_{sub}$ // sampling set, with size δN

1 /* initialization */
2 $\mathcal{R}_{sub} \leftarrow \varnothing$; // initial empty sampling set
3 $M \leftarrow int(\delta N)$; // the size of the final sampling set
4 /* get the first sample with pure entropy */
5 **foreach** $R_i \in \mathcal{R}$ **do**
6 $\quad$ compute the entropy $\hat{res}(R_i)$; // by Formula 4.10
7 $firstSample \leftarrow \underset{R_i}{\arg\max}\ \hat{res}(R_i)$
8 $\mathcal{R}_{sub}$.add($firstSample$);
9 /* get the following samples with mutual information */
10 **while** $|\mathcal{R}_{sub}| < M$ **do**
11 $\quad$ **foreach** $R_i \in \mathcal{R} - \mathcal{R}_{sub}$ **do**
12 $\quad\quad$ compute $\widetilde{res}(R_i; \mathcal{R}_{sub})$; // by Formula 4.11
13 $\quad$ $nextSample \leftarrow \underset{R_j}{\arg\max}\ \widetilde{res}(R_j)$
14 $\quad$ $\mathcal{R}_{sub}$.add($nextSample$);
15 **return** $\mathcal{R}_{sub}$;

These policies above give us a flavor of different techniques to employ for selecting sampling points from a series of mobile sensor readings.

4.3 ENERGY-EFFICIENT SEMANTIC ACTIVITY LEARNING

In the next case study, we focus our attention on another important dimension—energy-efficient sensing of *personal* activities, as opposed to a phenomenon around us. We review a methodology that recognizes typical locomotive state activities while reducing the power budget compared to traditional methods. Power consumption on mobile phones is a painful obstacle towards adoption of *continuous* sensing driven applications, e.g., continuously inferring individual's locomotive activities (such as "sit," "stand," or "walk") using the embedded accelerometer sensor. Previously in Chapter 3, we have covered techniques for semantic activity recognition from accelerometer streams. Hence, we would continue to use accelerometer-based activity recognition techniques for this chapter and demonstrate how to improve energy efficiency.

4.3.1   CLASSIFICATION ACCURACY VS. ENERGY CONSUMPTION

Let us recap the basic concepts to recognize semantic activities using accelerometer streams and understand its impact on energy consumption now. First, the incoming stream sampled at a certain *sampling frequency* is split into windows of fixed size. Thereafter, we identify key *classification features* needed to convert raw data in each window into a feature vector. Subsequently, several different learning algorithms (e.g., SVM, decision trees, etc.) are employed to classify the vector. Table 4.1 lists some of the commonly used features for activity recognition.

Table 4.1: Selected features used for activity recognition

Time Domain	Mean $(\bar{x}, \bar{y}, \bar{z})$, Magnitude $(\sqrt{x^2 + y^2 + z^2})$, Variance $\{var(x), var(y), var(z)\}$, Covariance $\{cov(x, y), cov(y, z), cov(x, z)\}$,
Frequency Domain	Energy $\left(\frac{\sum_{j=1}^{N}(m_j^2)}{N}\right)$, m_j is FFT component Entropy $\left(-\sum_{j=1}^{n}(p_j * \log(p_j))\right)$, p_j is FFT histogram

These features can be broadly classified into two categories—time-domain (F_{time}) and frequency domain (F_{freq}) features.

- **Time-domain Features (F_{time}).** These features are computed directly on the appropriate frames (e.g., 5 s, 10 s) of accelerometer streams; examples include the variance/mean of the frame as well as two-axis correlations.
- **Frequency-domain Features (F_{freq}).** Here, features such as entropy & energy are computed over frequency domain coefficients, which are first obtained by using FFT (or alternatives, such as wavelets) on each frame.

Several studies (e.g.,[91, 111]) have demonstrated that continuous activity recognition (applying on-board data processing over accelerometer and other sensor data streams) can rapidly drain the power on mobile devices. It is important to reduce the energy overheads of continuous mobile sensing. Let us focus on two independent parameters of the accelerometer-based activity recognition process: (a) the sensor sampling frequency and (b) the set and classes of features used in activity classification. Investigations (e.g., [83]) have established that these two parameters jointly influence a tradeoff between two important and mutually-conflicting objectives: (1) *Increase classification accuracy*: Increase in sampling frequency and a richer set of features both result in improved activity classification accuracy; (2) *Reduce energy overheads*: Conversely, reducing the sampling frequency, duty cycle and/or the set of features help to lower the energy overhead.

To reduce the energy overheads of accelerometer-based activity recognition, we first study the combined influence of these two parameters on the recognition accuracy, *separately for each distinct activity*. Many activity recognition techniques use the accelerometer features in an *activity-independent* fashion: the same set of features are used across all locomotive and postural states of the individual. We find that such an activity-independent use of features may not be optimal.

Different activities may be classified, with no or little loss in classification accuracy, by using different combinations of the ⟨*sampling frequency, classification features*⟩. The use of a lower sampling frequency or the computation of a smaller set of features should respectively impose a lower sensing and computational overhead. To verify this observation, we next report on the energy overheads and the classification accuracy for different combinations of *sampling frequency* (SF) and *classification feature* (CF) or the ⟨SF, CF⟩ tuples.

Classification Accuracy

To investigate the impact of different combinations of (SF, CF) domain on the accuracy of activity recognition, we study 10 specific activities: stand, slowWalk, sitRelax, sit, normalWalk, escalatorUp, escalatorDown, elevatorUp, elevatorDown, downStairs. To perform this study, we had four subjects perform each of the activities for a period of 5 min, resulting in a per subject duration of 50 min for 10 activities. Using these features, a J48 adaptive decision tree classifier is trained to build a personal classification model for each test subject.

The classification accuracy is then computed using 10-fold cross validation at different sampling frequencies and for different feature combinations. One aspect to note is that the training set was built using data collected using the maximum sampling frequency of 100 Hz. During testing (validation), the sampling frequency was varied to generate the test sets. The results of our classification conducted at different sampling frequencies is presented in Figures 4.5, 4.6, and 4.7. Kindly note that it is possible to collect training data at different sampling frequencies and verify how these perform to classify test sets at different frequencies. However, for on-phone deployments, as the training phase may be considered to be a one-time, offline activity, we assume that this one-time training data collection may be performed using the maximum sampling frequency.

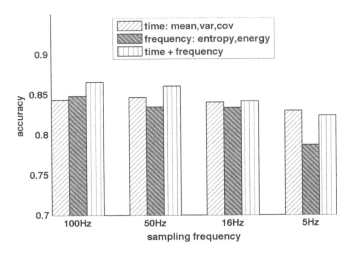

Figure 4.5: Accuracy at different ⟨SF, CF⟩ combinations.

Figure 4.5 depicts the classification accuracy averaged across 10 activities and 4 users as a function of sampling frequency and feature choices. We observe that (a) in many cases higher sampling frequency results in better accuracy and (b) using the combination of time and frequency domain features provides higher accuracy than using each feature type in isolation.

We next study the sensitivity of classification accuracy to different choices of $\langle SF, CF \rangle$ for each individual activity. Figures 4.6 and 4.7 show the average accuracy for a selected set of activities. Figure 4.6 considers only time-domain features, whereas Figure 4.7 considers both time and frequency domain features. We can see that the sensitivity of classification accuracy to different choices of $\langle SF, CF \rangle$ is clearly *activity-dependent*. For example, *sit* is correctly classified 95%+ of the time with $SF = 5Hz$ and using only F_{time} features, whereas the activities *stairs* and *escalators* benefits from using both F_{time} and F_{freq} features and a higher sampling frequency $SF = 100$ Hz.

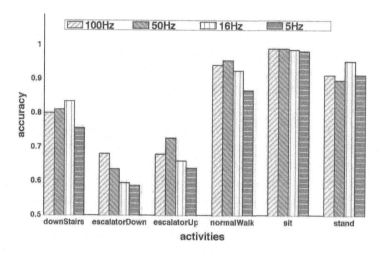

Figure 4.6: Activity-dependent accuracy using F_{time}.

The Energy Overhead

We study the dependency of the energy overhead on the combination of sampling frequency and the types of features computed (captured by the tuple $\langle SF, CF \rangle$).

Instead of considering all the possible combinations of the individual features listed in Table 4.1, we use *(i)* all *time-domain* features, *(ii)* all *frequency-domain* features or *(iii)* all *time+frequency domain* features. This approach is driven by the empirical observation that energy consumption is mainly impacted by inclusion or exclusion of a particular domain-specific feature group (time or frequency) and not an individual specific feature.

Figure 4.8 plots the energy consumption (in Joules) over a 2-h period, for different $\langle SF, CF \rangle$ combinations of two representative phones: Samsung Galaxy S2 and HTC Nexus 1.

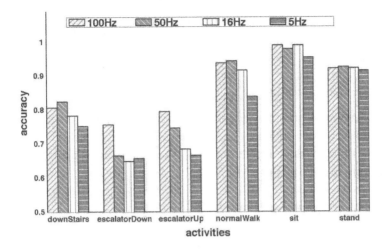

Figure 4.7: Activity-dependent accuracy using F_{time}+F_{freq}.

The Galaxy S2 and the Nexus 1 support sampling frequencies up to 100 Hz and 25 Hz, respectively. Note that the Android APIs only permits the sampling frequency between four discrete values ([5Hz, 16Hz, 50Hz, 100Hz]. To obtain the readings, we powered off the network interfaces and the display, and used the *PowerTutor* utility to measure the energy consumption.

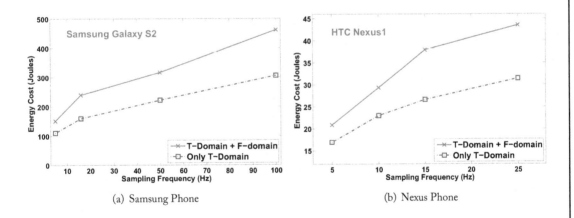

(a) Samsung Phone   (b) Nexus Phone

Figure 4.8: Energy consumption at different $\langle SF, CF \rangle$ combinations.

From these two energy profiles, we can observe the following points.

(1) The total energy overhead in continuous activity recognition clearly increases with sampling frequency.

(2) The increase in energy-overhead is nonlinear. Particularly, the *additional* energy overhead incurred by including frequency-domain features is a *logarithmic* function of the sampling frequency. This is due to the $O(n\log n)$ computational complexity, where n is the number of samples for a given frame of the required FFT operation needed during the computation of the frequency-domain features (F_{freq}).

Figure 4.8 demonstrates the tradeoffs between sampling frequency and features and motivates the need for techniques that exploit this tradeoff. For example, it is less expensive to utilize purely time-domain features at 50 Hz than using a combination of time and frequency domain features at the sampling frequency of 16 Hz.

Tradeoff between Energy and Accuracy

Table 4.2: Activity recognition accuracy using different $\langle SF, CF \rangle$ choices

Activity	Classification Accuracy							
	SamplingRate1 (100Hz)		SamplingRate2 (50Hz)		SamplingRate3 (16Hz)		SamplingRate4 (5Hz)	
	F_{time}	$F_{time}+F_{freq}$	F_{time}	$F_{time}+F_{freq}$	F_{time}	$F_{time}+F_{freq}$	F_{time}	$F_{time}+F_{freq}$
'stand'	0.9116	0.9203	0.8958	0.9244	0.9516	0.921	0.9123	0.9141
'slowWalk'	0.9379	0.935	0.9151	0.9069	0.9171	0.9064	0.8971	0.8486
'sitRelax'	0.9822	0.9821	0.9892	0.982	0.9856	0.9824	0.9717	0.9823
'sit'	0.989	0.989	0.9887	0.9783	0.9855	0.9889	0.9816	0.9535
'normalWalk'	0.9407	0.9364	0.9542	0.9424	0.9237	0.9154	0.8663	0.8386
'escalatorUp'	0.6786	0.7948	0.7265	0.7455	0.6592	0.6839	0.6378	0.6653
'escalatorDown'	0.6805	0.756	0.6356	0.6642	0.5947	0.6488	0.5868	0.6568
'elevatorUp'	0.7026	0.7606	0.7265	0.7863	0.7025	0.7224	0.7827	0.7596
'elevatorDown'	0.7353	0.7763	0.7648	0.8059	0.7669	0.7933	0.8056	0.7926
'downStairs'	0.8	0.8065	0.8097	0.8239	0.8344	0.7816	0.7559	0.7515
Consumed Power (J/hr)	152.75	230.75	110.05	158.4	79.95	119.9	55.35	75.8

Table 4.2 shows all the suitable combinations of $\langle SF, CF \rangle$ for all 10 activities and helps us understand the energy versus accuracy tradeoff better. As an example, for the activity *sit*, if we select $\langle 5Hz, F_{time} \rangle$ it gives us an accuracy of 98.16% at an energy consumption of 55.35 joules/h, but if we select $\langle 16Hz, F_{time} \rangle$ would have only gained marginally in terms of accuracy (98.55%), but have incurred a significantly higher energy consumption (80 joules/h).

We now use these results to create a smart operating configuration chart (Table 4.3) containing desired values for the feature states and sensor sampling states $\langle SF, CF \rangle$ for each activity that avoids both, a significant increase in the *energy consumed* and a steep drop in the *classification accuracy*. We use the following criteria for creating such a configuration chart.

- **Condition I (accuracy).** Given a desired minimal level of classification accuracy to be acc_{base} (e.g., 70%), we choose all states $\langle SF, CF \rangle$ that have registered an accuracy $acc_i \geq \Delta$, where $\Delta = max\{\delta \times \overline{accuracy}, acc_{base}\}$ for each activity. δ is a scaling coefficient and

$\overline{accuracy}$ is the average accuracy across all $\langle SF, CF \rangle$ for a given activity. For example, $\overline{accuracy}$ is the average of 8 values in each line in Table 4.2.

- **Condition II (energy).** Among the $\langle SF, CF \rangle$ states that satisfy *Condition I* for a given activity, we choose the state i that has the highest $\frac{accuracy_i}{power\ consumed_i}$.

Condition I ensures that we choose only from the acceptable $\langle ST, CF \rangle$ combinations, while Condition II chooses the most power-efficient combinations. We apply these two conditions for obtaining the operation configurations for the Samsung Galaxy S2 smartphone[5] using $\delta = 1$. The results of this procedure are shown in Table 4.3.

Table 4.3: Smart operating configuration for each activity (for our representative Samsung Galaxy S2)

activity	Smart choice	
'stand'	$16Hz$	F_{time}
'slowWalk'	$16Hz$	F_{time}
'sitRelax'	$5Hz$	$F_{time}+F_{freq}$
'sit'	$16Hz$	F_{time}
'normalWalk'	$16Hz$	F_{time}
'escalatorUp'	$50Hz$	F_{time}
'escalatorDown'	$100Hz$	$F_{time}+F_{freq}$
'elevatorUp'	$5Hz$	F_{time}
'elevatorDown'	$5Hz$	F_{time}
'downStairs'	$16Hz$	F_{time}

4.3.2 A3R—A METHODOLOGY FOR CONTINUOUS ADAPTIVE SAMPLING

Now we present an adaptive sampling approach that is able to dynamically choose a suitable sampling frequency and learning features to improve the energy consumption statistics during continuous sensing. This technique is called A3R (**A**daptive **A**ccelerometer-based **A**ctivity **R**ecognition)[168].

A3R assumes that an user is performing one of a set of N possible activities, denoted by $\mathcal{A} = \{A_0, A_1, A_2, \ldots, A_N\}$, where $A_i, (i = 1, \ldots, N)$ is either *sitting*, *standing*, or some such locomotive/postural state. A_0 represents the *unknown* activity state, where the activity classifier is unsure about the current activity. For each activity A_i, we assume the existence of an entry in the state table (similar to Table 4.3) that represents the best combination of sampling frequency and feature set for detecting A_i. We denote such an entry by $(\overline{SF_i}, \overline{CF_i})$.

---

[5]Note that the best operating point is device-specific. Different devices will have different energy curves and even different permitted sampling frequencies.

The A3R algorithm then works as follows. It starts off initially in the *unknown* state (A_0). In this state the accelerometer sampling rate is set to the highest frequency and the full combination of both time and frequency features is used. Each newly generated accelerometer frame is given to the classifier. The classifier returns a confidence vector $[p_1, p_2, \ldots, p_N]$ such that $\sum_{i=1}^{N} p_i = 1$, capturing the probability that the frame belongs to an activity state. Once the algorithm identifies with high confidence that the user is currently in a known activity state, say A_i, it switches the sampling frequency and the classification feature set to the corresponding optimal values $(\overline{SF_i}, \overline{CF_i})$.

The algorithm continues to use these values until it detects an episode where the classification confidence associated with the current ongoing activity drops below a certain threshold. At that point, A3R declares the user to have reverted to the unknown state A_0, switches back to the default case of high-frequency accelerometer sampling and uses the full set of classification features to newly re-establish a user's activity.

Specifically, A3R's state transition logic uses two parameters.

- W_{frame}. The number of consecutive frames (sliding windows) during which the confidence values are considered. The activity that has the highest average confidence value over recent W_{frame} frames is chosen as the best activity prediction in the current frame. This is done to eliminate erroneous/outlier frames or activity changes that last for short duration.

- Δ_{conf}. The threshold associated with detection of a state change. If the classifier's highest average confidence value smoothed over the last W_{frame} consecutive frames is above Δ_{conf}, A3R declares that the user is currently engaged in activity A_i. If not, A3R declares that the user's activity is unknown and switches back to A_0.

Algorithm 4.4 provides a concise description of the A3R algorithm. The algorithm effectively runs in a continuous loop, keeping track of the identified activity in the last W_{frame} activity frames to determine if the $\langle SF, CF \rangle$ combination still holds (i.e., the user is still in the current state) or if it should be switched to a new value (i.e., the user has transitioned to either a new activity or the unknown state).

Note that, in practice, A3R does not directly transition from one activity A_i to another activity $A_j (j \neq i)$ instantaneously. There is some delay as it takes some time steps for the next activity to gather confidence and remain above Δ_{conf}. More detailed performance evaluation and experimental results of this algorithm can be found in [168].

Algorithm 4.4: A3R Algorithm. The pseudocode describes the steady-state behavior of A3R. *Smooth_Confidence(t)* is the highest confidence value at frame t, across all N activities, averaged over the most recent W_{frame} activity frames, whereas *Smooth_Class(t)* is the index of the activity corresponding to this highest confidence value. A3R switches to the *unknown* state if the highest value of the smoothed confidence drops below Δ_{conf}.

---

1 $t \leftarrow 0; State \leftarrow A_0; State\_Table \leftarrow [A_0, A_1, \ldots, A_N];$

2 **while** *App_Running* **do**

3 $\quad Conf\_Vector(t) \leftarrow classify(Accel\_Frame(t));$

4 $\quad Smooth\_Vector(t) \leftarrow \frac{1}{W_{frame}} \sum_{i=t-W_{frame}}^{t} Conf\_Vector(i);$

5 $\quad Smooth\_Confidence(t) \leftarrow MAX(Smooth\_Vector(t));$

6 $\quad Smooth\_Class(t) \leftarrow Index(Smooth\_Confidence(t));$

7 $\quad$ **if** *Smooth_Confidence(t)* $< \Delta_{conf}$ **then**

8 $\quad\quad State \leftarrow State\_Table(0);$

9 $\quad\quad Choice \leftarrow \langle 100Hz, F_{time} + F_{freq} \rangle;$

10 $\quad$ **else**

11 $\quad\quad State \leftarrow State\_Table(Smooth\_Class);$

12 $\quad\quad Choice \leftarrow \langle \overline{SF_i}, \overline{CF_i} \rangle;$

13 $\quad t \leftarrow t + 1;$

4.4   CONCLUDING REMARKS

In this chapter we discussed key techniques for conserving energy while computing semantics from mobile sensor streams. The research area of energy-efficient mobile sensing is very broad. In this chapter, we focused on the key systemic steps that we need to be aware of, for writing energy-efficient algorithms on smartphone data streams.

The steps include sensor data sampling, statistical feature calculation, establishment of classification models. Further, we discussed two case studies. They cover optimal solutions for building adaptive sampling approaches for community-sensing, and how to adapt sampling frequency and feature selection for personalized sensing of activity semantics.

To design future energy-efficient semantic computation strategies from mobile sensors, we list a few interesting directions.

(1) **A full-functional activity adaptive approach.** In this chapter, we only discussed adaptive approaches that focused on dynamically making sound choices amongst sampling frequency and learning features. The energy cost of the estimating and using a model was less studies. Future research works could focus on finding the optimal combination of ⟨ sampling, feature, model ⟩ for each semantic activity.

(2) **Multi-sensory multi-user collaborative sensing.** The problem of optimal energy-efficient semantic computation over multiple users with various sensor types is NP-hard. Furthermore, the uncertainty of mobility in real-life applications makes the problem even more difficult. Certain mobility predication approaches and mobility patterns can be applied to decrease total energy consumption through collaboration.

CHAPTER 5

Conclusion

5.1 SUMMARY

With rapid adoption of mobile phone technologies, sensors and wearable devices have become commonplace today and have started to play an important role in the everyday life of users. The use of the smartphone as a device today is primarily on purpose, triggered by users to access standard services (call, SMS, chat log, etc.). However, a new paradigm is evolving where applications running on the phone passively collect user's context. This user context is used to improve the interaction of the user with non-traditional services like business applications (maps, health management, location services, traffic and transport, etc.). The businesses that provide such services have an advantage if they have access to user's context data. It enables them to better engage with the user. We are entering into a new era of mobile sensing where massive amounts heterogeneous sensing data are continuously generated from today's mobile devices. Over the last few years, researchers and developers have designed approaches to analyze this sensor data and to extract meaning from it. The sensors on the smartphone act as a "lens" to understand what the user is up to. It senses a user's personal context and the immersive environment of the user. We call this "semantics." In this book, we presented the key concepts of "semantics" and discussed various building-block techniques for computing such semantics from raw mobile sensing data collected using smartphone sensor data.

Mobile Sensing Evolution

The area of sensing has evolved considerably over the last 15–20 years. In the initial days, the focus was on conventional sensor networks. Thereafter, it moved to wireless sensor network (WSN), then to mobile WSNs where sensor nodes were mobilized following top-down policies. The area of people-centric sensing evolved around 2004–2005 due to the large-scale adoption of smartphones and the fundamental differences this "community of human-driven sensors" had with traditional mobile sensor networks. Apart from embedded sensors such as GPS, accelerometer, gyroscope, compass, camera, etc., a smartphone also acts as a "motherboard" for fixing other third-party sensors such as pluggable air quality, temperature, and pollution sensors. Going forward, as the number of body-worn sensors increase (e.g., Google glasses, smart watches), the smartphone is expected to also act as the hub for gathering context data from these sensors and act as the gateway to services on the Internet. At its core, these sensors essentially extract semantics about the user and the environment. Due to the vast amount of work in location trajectories using GPS and accelerometers for sensing motion semantics, this book focuses on the techniques for extracting semantics from these sensors. Subsequently, it dedicates a chapter that provides an overview of how to reduce energy expenditure that is needed to continuously run sensors on the phone.

Top-Down Semantic Modeling of Sensor Data

One fundamental direction researchers have taken is to provide top-down semantic and conceptual modeling of "what can be sensed." With an example of location and motion trajectories from GPS and accelerometer sensors, we focused on high-level semantic modeling of various levels of "meanings" that can be captured from these sensors. We expounded few fundamental concepts to model sensor semantics, i.e., *semantic trajectories*, *semantic activities*, and provided an indepth understanding of how to model location trajectories using ontologies. Ontologies play a key role in representing semantics of common knowledge of a given field, in particular in the study of semantic web.

Bottom-Up Semantic Computation of Sensor Data

Conventional semantic web technologies such as semantic reasoning provide us with a good view of the key concepts and standard models that may be used to achieve application interoperability. The main effort is to arrive at ontology standards that promote cross-application re-use of similar concepts. However, an important question is: how do we extract these semantics from the raw sensor data? This book provides details on data-driven techniques for computing semantics from raw sensor data. Raw sensor data is incrementally computed into different levels of data abstraction. For example, GPS data is first interpreted into different level semantics, i.e., as *spatio-temporal trajectories* after data preprocessing steps such as cleaning, compression, and segmentation. Thereafter, it is converted into *structured trajectories* composed of stops and move, and then into *semantic trajectories* by fusing the trajectories with Geographic Information Systems data such as road network, transportation modes, and point of interests. Similarly, for motion data generated by accelerometer sensor, researchers apply learning methods to first identify basic locomotive states such as sit, stand, walk, and jog, and then further deduce high-level semantic activities, e.g., work in office, meeting, cooking at home, dinner in restaurant. These statistical and machine learning approaches become critical techniques in computing semantics from the large-scale sensor trajectories generated in the mobile sensing era.

Energy Efficiency

The processing of sensor data from mobile devices consumes energy. Battery life being budgeted, this is a limitation that researchers must address. High data sampling rates on the mobile devices may cause the battery to drain faster than usual, causing undesirable user experiences. This book illustrates a few key energy-efficient computation approaches.

We present a summary of how energy efficiency may be achieved at different computation layers, i.e., data sampling, feature computation, classification models. We present a two-tier model for adapting sampling frequencies while sensing environmental phenomenon using community-carried sensors. We also present another work that adapts sampling frequency and feature sets jointly, to reduce energy consumption while maintaining the same quality of inference of semantic activities from accelerometer streams.

5.2 CHALLENGES AND OPPORTUNITIES

The area of extracting meaningful user context from smartphone sensors is relatively new. There are many challenges and opportunities for further studies around extracting semantics from smartphone sensor data. Researchers are working on other areas apart from the ones we could cover in this book. A few notable directions are illustrated below.

"Soft" Sensors and Crowdsensing

In this book, we particularly concentrated on three most well-known mobile sensor types and illustrated methods for modeling and computing semantics from these representative sensors, i.e., location sensors like GPS, motion sensors like accelerometer, and third-party sensors like environmental measurers. In addition to GPS and accelerometer, Chapter 1 discusses use of other embedded phone sensors, such as Bluetooth, Light, Microphone, Camera, etc. Bluetooth sensors can be used to infer the semantics social dynamics around users, e.g., detecting user's social behaviors [119], estimating crowd density [117], etc. Light sensors can be applied for inferring contexts such as "indoor vs. outdoor," and "day vs. night" or to improve indoor localization algorithms [7]. A typical use-case is of automatic image tagging while taking photos with smartphone [133]. Additional sensors are used to tag the picture with user's context (where it was taken, sound level, crowd level, etc.). Acoustic sensors (microphones) have been used for identifying speakers and detecting the roles such as group members and leadership [105], and acoustics is also applicable to deduce noise map in the environment [138].

In addition to these physical sensors, Chapter 1 briefly addressed the use of other "soft sensors" such as app usages, mobile internet access, phone calls, etc., and how user context may be extracted from these. Data collected by users on their phones (either passively by sensors or actively with user inputs) collectively can be very valuable in providing services back to the crowd. Taking advantage of this model, different crowdsensing applications (e.g., GeoCha, Ushahidi, GasBuddy) are being launched. How to efficiently process and extract relevant information from such inherently unreliable, noisy data from user inputs via different smartphone sensors is a challenge.

Privacy and Security

Privacy and security form an important concern for developing and deploying mobile sensing applications. The sensors contain valuable personal context and sharing this data with applications gives rise to these concerns. Therefore, researchers are aware of the importance to build privacy-aware mobile sensing systems. For example, Cornelius et al. in [35] designs AnonySense, a privacy preserving architecture for collecting sensing data from personal mobile devices through collaborative and opportunistic sensing. In AnonySense, a trust model and specific security properties are designed for protecting the privacy of participants while allowing their devices to reliably contribute high-quality data to these large-scale sensing applications.

Semantic trajectories generated from mobile users can enable many advanced location based services, such as location based queries, social recommendations, monitoring traffic, and real-time navigation. However, these also raise serious privacy worries, such as the leakage of sensitive information such as individual's health status (e.g., going to an ADIS hospital), political and religious afflictions (e.g., periodically visit a dedicated church), etc. Ghiiniitta et al. recently summarized key techniques for location privacy protection [61]. In addition to location, many other sensors like Bluetooth proximity and acoustic characteristics also convey a lot of personal context. Any application or system attempting to tap into these sensors need to address privacy and security concerns of users.

The focus of this book is on the algorithms to extract semantics and not techniques for data obfuscation or security mechanisms. Readers interested in privacy and security models on top of such data are encouraged to refer to [58][61]. This forms another significant area of further work.

Integration with Cloud Infrastructure

With the growth of cloud computing technologies, the combined infrastructure of a powerful cloud computing back-end for processing peta-scale data streams and mobile sensors becomes an important topic in both academia and industry. The advantage of the cloud platform lies in not only providing higher compute resources, but also as a coordination platform. This allows industries and governments to start imagining the deployment of large-scale platforms for collecting city-scale sensing data. For example, Flinn et al. [53] presents a new topic of "cyber foraging" that lies at the intersection of mobile computing and cloud computing. The main focus of cyber foraging is on dynamic partitioning of application functionality and executing the computation on the cloud on behalf of mobile users. A detailed survey about mobile cloud computing can be found in [40]. Recently, a new keyword "physical analytics" starts to gain momentum.[1] The workshop topic states: "While huge strides have been made in online analytics to extract a wealth of information from peoples' online activities, corresponding work in the physical context—which we term as Physical Analytics—is relatively nascent and scattered." The collaboration of a cloud platform along with community-carried sensors may improve the scale and capability at which users and our environments can be sensed. This is an area that should see future work.

Integration with Mobile Social Network

Social networks are a reality today. User context sensed using smartphone data can be easily shared with our social network. In fact, sharing of pictures, status updates, etc. in a way are instances of user's context being shared with our online network. Users can be easily connected in using mobile devices through social networking platforms, e.g., Facebook, Twitter, as well as many new

---

[1]http://www.sigmobile.org/mobisys/2014/workshops/physicalanalytics/

and emerging crowd-sourcing applications, such as GeoCha, Ushahidi, GasBuddy, and Waze. On one side, mobile sensing researches focus on sensing of physical sensors such as GPS for location and accelerometer for motion. On the other side, these mobile sensor networks foster a new scope of "soft'" sensing of social phenomena using mobile devices—e.g., real-time crowd-sourcing about weather information and traffic conditions reported by end users. Mobile social network plays an active role in generating massive "soft" sensor data on people's daily mobile devices such as smartphones and tablets. Mining joint semantics from phone physical sensor data and soft sensing data from mobile social networks will be an interesting and challenging new topic.

Bibliography

[1] S. Abdullah, N. D. Lane, and T. Choudhury. Towards Population Scale Activity Recognition: A Framework for Handling Data Diversity. In *AAAI*, 2012. 84

[2] K. Aberer, S. Sathe, D. Chakraborty, A. Martinoli, G. Barrenetxea, B. Faltings, and L. Thiele. OpenSense: open community driven sensing of environment. In *GIS-IWGS*, pages 39–42, 2010. DOI: 10.1145/1878500.1878509. 5, 87, 88, 94

[3] I. F. Akyildiz, W. Su, Y. Sankarasubramaniam, and E. Cayirci. A survey on Sensor Networks. *IEEE Communications magazine*, 40(8):102–114, 2002. DOI: 10.1109/MCOM.2002.1024422. 1

[4] I. F. Akyildiz, W. Su, Y. Sankarasubramaniam, and E. Cayirci. Wireless sensor networks: a survey. *Computer networks*, 38(4):393–422, 2002. DOI: 10.1016/S1389-1286(01)00302-4. 1

[5] L. O. Alvares, V. Bogorny, B. Kuijpers, J. Macedo, B. Moelans, and A. Vaisman. A Model for Enriching Trajectories with Semantic Geographical Information. In *GIS*, page 22, 2007. DOI: 10.1145/1341012.1341041. 16, 45

[6] G. Andrienko, N. Andrienko, and M. Heurich. An Event-Based Conceptual Model for Context-Aware Movement Analysis. *International Journal Geographical Information Science*, 25(9):1347–1370, 2011. DOI: 10.1080/13658816.2011.556120. 25

[7] M. Azizyan and R. R. Choudhury. SurroundSense: mobile phone localization using ambient sound and light. *ACM SIGMOBILE Mobile Computing and Communications Review*, 13(1):69–72, 2009. DOI: 10.1145/1558590.1558605. 4, 52, 111

[8] P. Bahl and V. N. Padmanabhan. RADAR: An in-building RF-based user location and tracking system. In *INFOCOM*, volume 2, pages 775–784, 2000. DOI: 10.1109/INFCOM.2000.832252. 13

[9] L. Bao and S. Intille. Activity Recognition from User-Annotated Acceleration Data. In *Pervasive*, pages 1–17, 2004. DOI: 10.1007/978-3-540-24646-6_1. 53, 61, 71

[10] N. Beckmann, H.-P. Kriegel, R. Schneider, and B. Seeger. The R*-Tree: An Efficient and Robust Access Method for Points and Rectangles. *SIGMOD Record*, 19(2):322–331, 1990. DOI: 10.1145/93605.98741. 40, 42

[11] R. Bellman. On the Approximation of Curves by Line Segments Using Dynamic Programming. *Communications of the ACM*, 4(6):284, 1961. DOI: 10.1145/366573.366611. 94

[12] T. Berners-Lee, J. Hendler, O. Lassila, et al. The Semantic Web. *Scientific american*, 284(5):28–37, 2001. DOI: 10.1038/scientificamerican0501-34. 8

[13] U. Blanke and B. Schiele. Daily routine recognition through activity spotting. In *Location and Context Awareness*, pages 192–206. Springer, 2009. DOI: 10.1007/978-3-642-01721-6_12. 9

[14] U. Blanke and B. Schiele. Remember and Transfer what you have Learned - Recognizing Composite Activities based on Activity Spotting. In *ISWC*, pages 1–8, 2010. DOI: 10.1109/ISWC.2010.5665869. 77

[15] D. M. Blei, A. Y. Ng, and M. I. Jordan. Latent Dirichlet Allocation. *Journal of Machine Learning Research*, 3:993–1022, 2003. 77

[16] A. Borgida, V. K. Chaudhri, P. Giorgini, and E. S. K. Yu, editors. *Conceptual Modeling: Foundations and Applications - Essays in Honor of John Mylopoulos*, volume 5600, 2009. DOI: 10.1007/978-3-642-02463-4. 8

[17] M. Boulos, B. Resch, D. Crowley, J. Breslin, G. Sohn, R. Burtner, W. A. Pike, E. Jezierski, and K. Chuang. Crowdsourcing, citizen sensing and sensor web technologies for public and environmental health surveillance and crisis management: trends, ogc standards and application examples. *International journal of health geographics*, 10(1):67, 2011. DOI: 10.1186/1476-072X-10-67. 5

[18] S. Brakatsoulas, D. Pfoser, R. Salas, and C. Wenk. On Map-Matching Vehicle Tracking Data. In *VLDB*, pages 853–864, 2005. 28, 42

[19] T. Brinkhoff, H.-P. Kriegel, and B. Seeger. Efficient Processing of Spatial Joins using R-Trees. In *SIGMOD*, pages 237–246, 1993. DOI: 10.1145/170036.170075. 40

[20] I. Burbey and T. L. Martin. A survey on predicting personal mobility. *International Journal of Pervasive Computing and Communications*, 8(1):5–22, 2012. DOI: 10.1108/17427371211221063. 9

[21] J. A. Burke, D. Estrin, M. Hansen, A. Parker, N. Ramanathan, S. Reddy, and M. B. Srivastava. Participatory sensing. In *SenSys World Sensor Web Workshop*, 2006. 6

[22] T. Camp, J. Boleng, and V. Davies. A survey of mobility models for ad hoc network research. *Wireless communications and mobile computing*, 2(5):483–502, 2002. DOI: 10.1002/wcm.72. 1

[23] A. Campbell, S. Eisenman, N. Lane, E. Miluzzo, R. Peterson, H. Lu, X. Zheng, M. Musolesi, K. Fodor, and G.-S. Ahn. The Rise of People-Centric Sensing. *IEEE Internet Computing*, 12:12–21, 2008. DOI: 10.1109/MIC.2008.90. 2

[24] A. Campbell, S. B. Eisenman, N. D. Lane, E. Miluzzo, and R. A. Peterson. People-centric urban sensing. In *Proceedings of the 2nd annual international workshop on Wireless internet*, page 18, 2006. DOI: 10.1145/1234161.1234179. 2, 6

[25] M. Cardei and J. Wu. Energy-efficient coverage problems in wireless ad-hoc sensor networks. *Computer communications*, 29(4):413–420, 2006. DOI: 10.1016/j.comcom.2004.12.025. 1

[26] A. Carroll and G. Heiser. An analysis of power consumption in a smartphone. In *USENIX annual technical conference*, pages 21–21, 2010. 85, 86

[27] S. Cartier, S. Sathe, D. Chakraborty, and K. Aberer. ConDense: Managing data in community-driven mobile geosensor networks. In *SECON*, pages 515–523, 2012. DOI: 10.1109/SECON.2012.6275820. 88

[28] H. Chan, M. Yang, H. Zheng, H. Wang, R. Sterritt, S. McClean, and R. Mayagoitia. Machine learning and statistical approaches to assessing gait patterns of younger and older healthy adults climbing stairs. In *ICNC*, volume 1, pages 588–592, 2011. DOI: 10.1109/ICNC.2011.6022097. 54

[29] S. Chan, B. Ranalli, C. Kuo, K. Intrator, and C. Atencio. Humans as sensors: Fusion of participatory mechanisms and computational innovations to monitor climate change and its consequences. In *ACCORD volume*, pages 1–19. Institute for Geoinformatics, 2012. 5

[30] C.-C. Chang and C.-J. Lin. Libsvm: A library for support vector machines. *ACM TIST*, 2(3):27, 2011. DOI: 10.1145/1961189.1961199. 79

[31] S. Chen, C. S. Jensen, and D. Lin. A Benchmark for Evaluating Moving Object Indexes. In *VLDB*, volume 1(2), pages 1574–1585, 2008. 15

[32] H. Cheng, X. Yan, J. Han, and C. Hsu. Discriminative Frequent Pattern Analysis for Effective Classification. In *ICDE*, pages 716–725, 2007. DOI: 10.1109/ICDE.2007.367917. 75

[33] T. Choudhury, S. Consolvo, B. Harrison, J. Hightower, A. LaMarca, L. LeGrand, A. Rahimi, A. Rea, G. Bordello, and B. Hemingway. The mobile sensing platform: An embedded activity recognition system. *IEEE Pervasive Computing*, 7(2):32–41, 2008. DOI: 10.1109/MPRV.2008.39. 69

[34] D. Chu, N. D. Lane, T. T.-T. Lai, C. Pang, X. Meng, Q. Guo, F. Li, and F. Zhao. Balancing energy, latency and accuracy for mobile sensor data classification. In *SenSys*, page 54, 2011. DOI: 10.1145/2070942.2070949. 62, 89

[35] C. Cornelius, A. Kapadia, D. Kotz, D. Peebles, M. Shin, and N. Triandopoulos. Anonysense: privacy-aware people-centric sensing. In *MobiSys*, pages 211–224, 2008. DOI: 10.1145/1378600.1378624. 111

[36] D. Cuff, M. Hansen, and J. Kang. Urban sensing: out of the woods. *Communications of the ACM*, 51(3):24–33, 2008. DOI: 10.1145/1325555.1325562. 6

[37] H. Dejnabadi, B. M. Jolles, and K. Aminian. A new approach to accurate measurement of uniaxial joint angles based on a combination of accelerometers and gyroscopes. *IEEE Transactions on Biomedical Engineering*, 52(8):1478–1484, 2005. DOI: 10.1109/TBME.2005.851475. 3

[38] S. Dhillon and K. Chakrabarty. Sensor Placement for Effective Coverage and Surveillance in Distributed Sensor Networks. In *WCNC*, pages 1609–1614, 2003. 88

[39] S. S. Dhillon and K. Chakrabarty. *Sensor placement for effective coverage and surveillance in distributed sensor networks*, volume 3. IEEE, 2003. DOI: 10.1109/WCNC.2003.1200627. 1

[40] H. T. Dinh, C. Lee, D. Niyato, and P. Wang. A survey of mobile cloud computing: architecture, applications, and approaches. *Wireless Communications and Mobile Computing*, 2011. DOI: 10.1016/j.future.2012.05.023. 112

[41] T. M. T. Do and D. Gatica-Perez. Groupus: Smartphone proximity data and human interaction type mining. In *ISWC*, pages 21–28. IEEE, 2011. DOI: 10.1109/ISWC.2011.28. 3

[42] C. du Mouza and P. Rigaux. Mobility Patterns. *GeoInformatica*, 9(4):297–319, 2005. DOI: 10.1007/s10707-005-4574-9. 16

[43] C. Düntgen, T. Behr, and R. H. Güting. BerlinMOD: A Benchmark for Moving Object Databases. *The VLDB Journal*, 18(4), 2007. DOI: 10.1007/s00778-009-0142-5. 15

[44] J. Eberle and G. P. Perrucci. Energy measurements campaign for positioning methods on state-of-the-art smartphones. In *CCNC*, pages 937–941. IEEE, 2011. DOI: 10.1109/CCNC.2011.5766645. 85, 86

[45] H. Edner, K. Fredriksson, A. Sunesson, S. Svanberg, L. Uéus, and W. Wendt. Mobile remote sensing system for atmospheric monitoring. *Applied optics*, 26(19):4330–4338, 1987. DOI: 10.1364/AO.26.004330. 1

[46] N. El Mawass and E. Kanjo. A supermarket stress map. In *Ubicomp*, pages 1043–1046, 2013. DOI: 10.1145/2494091.2496017. 52

[47] J. Eriksson, L. Girod, B. Hull, R. Newton, S. Madden, and H. Balakrishnan. The pothole patrol: using a mobile sensor network for road surface monitoring. In *MobiSys*, pages 29–39, 2008. DOI: 10.1145/1378600.1378605. 1

[48] M. Ermes, J. Parkka, and L. Cluitmans. Advancing from offline to online activity recognition with wearable sensors. In *EMBS*, pages 4451–4454, 2008. DOI: 10.1109/IEMBS.2008.4650199. 54

[49] M. Erwig, R. H. Güting, M. Schneider, and M. Vazirgiannis. Spatio-Temporal Data Types: An Approach to Modeling and Querying Moving Objects in Databases. *Geoinformatica*, 3(3):269–296, 1999. DOI: 10.1023/A:1009805532638. 20

[50] M. Ester, H.-P. Kriegel, J. Sander, and X. Xu. A Density-Based Algorithm for Discovering Clusters in Large Spatial Databases with Noise. In *KDD*, pages 226–231, 1996. 36

[51] M. Faulkner, M. Olson, R. Chandy, J. Krause, K. M. Chandy, and A. Krause. The next big one: Detecting earthquakes and other rare events from community-based sensors, 2011. 5

[52] D. Ferreira, A. K. Dey, and V. Kostakos. Understanding human-smartphone concerns: a study of battery life. In *Pervasive Computing*, pages 19–33. Springer, 2011. DOI: 10.1007/978-3-642-21726-5_2. 85

[53] J. Flinn. *Cyber Foraging: Bridging Mobile and Cloud Computing*. Synthesis Lectures on Mobiel and Pervasive Computing. Morgan & Claypool, 2012. 112

[54] F. Foerster, M. Smeja, and J. Fahrenberg. Detection of posture and motion by accelerometry: a validation study in ambulatory monitoring. *Computers in Human Behavior*, 15(5):571–583, 1999. DOI: 10.1016/S0747-5632(99)00037-0. 61

[55] G. D. Forney. The Viterbi Algorithm. *Proceedings of the IEEE*, 61(3):268–278, 1973. DOI: 10.1109/PROC.1973.9030. 48

[56] E. Frentzos. *Trajectory Data Management in Moving Object Databases*. PhD thesis, University of Piraeus, 2008. 15, 27, 28

[57] Y. Freund and R. E. Schapire. A desicion-theoretic generalization of on-line learning and an application to boosting. In *Computational learning theory*, pages 23–37. Springer, 1995. DOI: 10.1007/3-540-59119-2_166. 65

[58] B. Fung, K. Wang, R. Chen, and P. S. Yu. Privacy-preserving data publishing: A survey of recent developments. *ACM Computing Surveys*, 42(4):14, 2010. DOI: 10.1145/1749603.1749605. 10, 112

[59] F. Fusier, V. Valentin, F. Brémond, M. Thonnat, M. Borg, D. Thirde, and J. Ferryman. Video understanding for complex activity recognition. *Machine Vision and Applications*, 18(3-4):167–188, 2007. DOI: 10.1007/s00138-006-0054-y. 54

[60] R. K. Ganti, F. Ye, and H. Lei. Mobile crowdsensing: Current state and future challenges. *IEEE Communications Magazine*, 49(11):32–39, 2011. DOI: 10.1109/M-COM.2011.6069707. 6

[61] G. Ghiiniitta. *Privacy for Location-Based Services*. Morgan & Claypool, 2013. DOI: 10.2200/S00485ED1V01Y201303SPT004. 112

[62] F. Giannotti and D. Pedreschi. *Mobility, Data Mining and Privacy, Geographic Knowledge Discovery*. Springer-Verlag, 2008. DOI: 10.1007/978-3-540-75177-9. 15, 16, 45

[63] J. Goldman, K. Shilton, J. Burke, D. Estrin, M. Hansen, N. Ramanathan, S. Reddy, V. Samanta, M. Srivastava, and R. West. Participatory sensing: A citizen-powered approach to illuminating the patterns that shape our world. *Foresight & Governance Project, White Paper*, 2009. 6

[64] L. I. Gómez and A. A. Vaisman. Efficient Constraint Evaluation in Categorical Sequential Pattern Mining for Trajectory Databases. In *EDBT*, pages 541–552, 2009. DOI: 10.1145/1516360.1516423. 16

[65] T. R. Gruber. Toward principles for the design of ontologies used for knowledge sharing? *International journal of human-computer studies*, 43(5):907–928, 1995. DOI: 10.1006/i-jhc.1995.1081. 8

[66] T. Gu, Z. Wu, X. Tao, H. K. Pung, and J. Lu. epSICAR: An Emerging Patterns based Approach to Sequential, Interleaved and Concurrent Activity Recognition. In *PerCom*, pages 1–9, 2009. DOI: 10.1109/PERCOM.2009.4912776. 9, 54, 61, 71, 84

[67] R. H. Güting, M. H. Böhlen, M. Erwig, C. S. Jensen, N. A. Lorentzos, M. Schneider, and M. Vazirgiannis. A Foundation for Representing and Querying Moving Objects. *ACM Transactions on Database Systems*, 25(1):1–42, 2000. DOI: 10.1145/352958.352963. 15, 20

[68] R. H. Güting, V. T. de Almeida, D. Ansorge, T. Behr, Z. Ding, T. Höse, F. Hoffmann, M. Spiekermann, and U. Telle. SECONDO: An Extensible DBMS Platform for Research Prototyping and Teaching. In *ICDE*, pages 1115–1116, 2005. 14

[69] R. H. Güting, V. T. de Almeida, and Z. Ding. Modeling and querying moving objects in networks. *The VLDB Journal*, 15(2):165–190, 2006. DOI: 10.1007/s00778-005-0152-x. 28

[70] R. H. Güting and M. Schneider. Realm-Based Spatial Data Types: The ROSE Algebra. *VLDB Journal*, 4:243–286, 1995. DOI: 10.1007/BF01237921. 39

[71] R. H. Güting and M. Schneider. *Moving Objects Databases*. Morgan Kaufmann, 2005. 16

[72] A. Guttman. R-Trees: A Dynamic Index Structure for Spatial Searching. In *SIGMOD*, pages 47–57, 1984. DOI: 10.1145/971697.602266. 14

[73] M. Halkidi, Y. Batistakis, and M. Vazirgiannis. On clustering validation techniques. *J. Intell. Inf. Syst.*, 17(2-3):107–145, 2001. DOI: 10.1023/A:1012801612483. 66

[74] J. Han, M. Kamber, and J. Pei. *Data mining: concepts and techniques*. Morgan kaufmann, 2006. 6

[75] J. Han, J.-G. Lee, H. Gonzalez, and X. Li. Mining Massive RFID, Trajectory, and Traffic Data Sets (Tutorial). In *KDD*, 2008. DOI: 10.1145/1401890.1551566. 15

[76] D. Hasenfratz, O. Saukh, S. Sturzenegger, and L. Thiele. Participatory Air Pollution Monitoring Using Smartphones. In *1st International Workshop on Mobile Sensing: From Smartphones and Wearables to Big Data*, 2012. 89

[77] J. Himberg, K. Korpiaho, H. Mannila, J. Tikanmäki, and H. Toivonen. Time Series Segmentation for Context Recognition in Mobile Devices. In *ICDM*, pages 203–210, 2001. DOI: 10.1109/ICDM.2001.989520. 94

[78] A. Howard, M. J. Matarić, and G. S. Sukhatme. Mobile sensor network deployment using potential fields: A distributed, scalable solution to the area coverage problem. In *Distributed Autonomous Robotic Systems 5*, pages 299–308. Springer, 2002. DOI: 10.1007/978-4-431-65941-9_30. 1

[79] J. Howe. The rise of crowdsourcing. *Wired magazine*, 14(6):1–4, 2006. 6

[80] T. Huynh, M. Fritz, and B. Schiele. Discovery of activity patterns using topic models. In *Ubicomp*, pages 10–19, 2008. DOI: 10.1145/1409635.1409638. 77

[81] H. Ishida, K.-i. Suetsugu, T. Nakamoto, and T. Moriizumi. Study of autonomous mobile sensing system for localization of odor source using gas sensors and anemometric sensors. *Sensors and Actuators A: Physical*, 45(2):153–157, 1994. DOI: 10.1016/0924-4247(94)00829-9. 1

[82] C. S. Jensen, D. Tiesyte, and N. Tradisauskas. The COST Benchmark-Comparison and Evaluation of Spatio-temporal Indexes. In *DASFAA*, pages 125–140, 2006. DOI: 10.1007/11733836_11. 15

[83] H. Junker, P. Lukowicz, and G. Troster. Sampling frequency, signal resolution and the accuracy of wearable context recognition systems. In *ISWC*, volume 1, pages 176–177. IEEE, 2004. DOI: 10.1109/ISWC.2004.38. 62, 100

[84] M. Kantardzic. *Data mining: concepts, models, methods, and algorithms*. Wiley-IEEE Press, 2011. DOI: 10.1002/9781118029145. 6

[85] A. Kapadia, D. Kotz, and N. Triandopoulos. Opportunistic sensing: Security challenges for the new paradigm. In *COMSNETS*, pages 1–10. IEEE, 2009. DOI: 10.1109/COMSNETS.2009.4808850. 6

[86] E. Keogh, S. Chu, D. Hart, and M. Pazzani. *Segmenting Time Series: A Survey and Novel Approach*, pages 1–22. World Scientific Publishing, 2004. 32, 95

[87] A. M. Khan, Y.-K. Lee, S. Y. Lee, and T.-S. Kim. A triaxial accelerometer-based physical-activity recognition via augmented-signal features and a hierarchical recognizer. *IEEE Transactions on Information Technology in Biomedicine*, 14(5):1166–1172, 2010. DOI: 10.1109/TITB.2010.2051955. 54

[88] E. Kim, S. Helal, and D. Cook. Human activity recognition and pattern discovery. *IEEE Pervasive Computing*, 9(1):48–53, 2010. DOI: 10.1109/MPRV.2010.7. 9

[89] A. Kittur, E. H. Chi, and B. Suh. Crowdsourcing user studies with mechanical turk. In *SIGCHI*, pages 453–456, 2008. DOI: 10.1145/1357054.1357127. 6

[90] A. Krause, E. Horvitz, A. Kansal, and F. Zhao. Toward community sensing. In *ISPN*, pages 481–492. IEEE Computer Society, 2008. DOI: 10.1109/IPSN.2008.37. 6

[91] A. Krause, M. Ihmig, E. Rankin, D. Leong, S. Gupta, D. Siewiorek, A. Smailagic, M. Deisher, and U. Sengupta. Trading off prediction accuracy and power consumption for context-aware wearable computing. In *ISWC*, pages 20–26. IEEE, 2005. DOI: 10.1109/ISWC.2005.52. 61, 100

[92] A. Krause, A. Singh, and C. Guestrin. Near-Optimal Sensor Placements in Gaussian Processes: Theory, Efficient Algorithms and Empirical Studies. *Journal of Machine Learning Research*, 9:235–284, 2008. DOI: 10.1145/1390681.1390689. 88, 98

[93] B. Kuijpers and W. Othman. Trajectory Databases: Data Models, Uncertainty and Complete Query Languages. In *ICDT*, pages 224–238, 2007. DOI: 10.1007/11965893_16. 16, 20

[94] K. Kunze and P. Lukowicz. Dealing with sensor displacement in motion-based on-body activity recognition systems. In *Ubicomp*, pages 20–29. ACM, 2008. DOI: 10.1145/1409635.1409639. 52

[95] N. D. Lane, S. B. Eisenman, M. Musolesi, E. Miluzzo, and A. T. Campbell. Urban sensing systems: opportunistic or participatory? In *HotMobile*, pages 11–16. ACM, 2008. DOI: 10.1145/1411759.1411763. 6

[96] N. D. Lane, E. Miluzzo, H. Lu, D. Peebles, T. Choudhury, and A. T. Campbell. A survey of mobile phone sensing. *IEEE Communications Magazine*, 48(9):140–150, 2010. DOI: 10.1109/MCOM.2010.5560598. 2, 4

[97] S. Lee, C. Min, C. Yoo, and J. Song. Understanding customer malling behavior in an urban shopping mall using smartphones. In *Ubicomp*, pages 901–910. ACM, 2013. DOI: 10.1145/2494091.2497344. 52

[98] J. Leppanen and A. Eronen. Accelerometer-based activity recognition on a mobile phone using cepstral features and quantized gmms. In *ICASSP*, pages 3487–3491. IEEE, 2013. DOI: 10.1109/ICASSP.2013.6638306. 54, 61

[99] Y. Li, M. Chen, Q. Li, and W. Zhang. Enabling multilevel trust in privacy preserving data mining. *IEEE Transactions on Knowledge and Data Engineering*, 24(9):1598–1612, 2012. DOI: 10.1109/TKDE.2011.124. 10

[100] Z. Li, M. Ji, J.-G. Lee, L.-A. Tang, Y. Yu, J. Han, and R. Kays. MoveMine: Mining Moving Object Databases. In *SIGMOD*, pages 1203–1206, 2010. 15

[101] F. Lin and P. Chiu. A Near-Optimal Sensor Placement Algorithm to Achieve Complete Coverage-Discrimination in Sensor Networks. *IEEE Communications Letters*, 9(1):43 – 45, 2005. DOI: 10.1109/LCOMM.2005.1375236. 88

[102] S. Lin, B. Arai, D. Gunopulos, and G. Das. Region Sampling: Continuous Adaptive Sampling on Sensor Networks. In *ICDE*, pages 794–803, 2008. DOI: 10.1109/ICDE.2008.4497488. 88

[103] B. Lorenz, H. J. Ohlbach, and L. Yang. Ontology of Transportation Networks, 2005. 21

[104] H. Lu, A. J. B. Brush, B. Priyantha, A. K. Karlson, and J. Liu. Speakersense: Energy efficient unobtrusive speaker identification on mobile phones. In *Pervasive*, pages 188–205, 2011. DOI: 10.1007/978-3-642-21726-5_12. 61, 87

[105] H. Lu, W. Pan, N. D. Lane, T. Choudhury, and A. T. Campbell. Soundsense: scalable sound sensing for people-centric applications on mobile phones. In *MobiSys*, pages 165–178. ACM, 2009. DOI: 10.1145/1555816.1555834. 111

[106] H. Lu, J. Yang, J. Liu, N. Lane, T. Choudhury, and A. Campbell. The Jigsaw Continuous Sensing Engine for Mobile Phone Applications. In *Sensys*, pages 71–84, 2010. DOI: 10.1145/1869983.1869992. 69

[107] H. Luinge and P. H. Veltink. Measuring orientation of human body segments using miniature gyroscopes and accelerometers. *Medical and Biological Engineering and computing*, 43(2):273–282, 2005. DOI: 10.1007/BF02345966. 3

[108] S. Mann. Wearable computing: A first step toward personal imaging. *Computer*, 30(2):25–32, 1997. DOI: 10.1109/2.566147. 1

[109] D. Mark, M. Egenhofer, L. Bian, K. Hornsby, P. Rogerson, and J. Vena. Spatio-temporal GIS Analysis for Environmental Health Using Geospatial Lifelines. In *GEOMED*, 1999. 20

[110] N. Meratnia and R. A. de By. Spatiotemporal Compression Techniques for Moving Point Objects. In *EDBT*, pages 765–782, 2004. DOI: 10.1007/978-3-540-24741-8_44. 29

[111] E. Miluzzo, N. Lane, K. Fodor, R. Peterson, H. Lu, M. Musolesi, S. Eisenman, X. Zheng, and A. Campbell. Sensing Meets Mobile Social Networks: the Design, Implementation and Evaluation of the Cenceme Application. In *Sensys*, pages 337–350, 2008. DOI: 10.1145/1460412.1460445. 61, 100

[112] D. Mizell. Using gravity to estimate accelerometer orientation. In *ISWC '03*, page 17, 2003. DOI: 10.1109/ISWC.2003.1241424. 59

[113] M. Mokbel, X. Xiong, W. G. Aref, S. E. Hambrusch, S. Prabhakar, and M. A. Hammad. PLACE: A Query Processor for Handling Real-time Spatio-temporal Data Streams. In *VLDB*, pages 1377–1380, 2004. 14

[114] M. F. Mokbel, T. M. Ghanem, and W. G. Aref. Spatio-temporal access methods. *IEEE Data Eng. Bull.*, 26(2):40–49, 2003. 15

[115] J. Myllymaki and J. H. Kaufman. DynaMark: A Benchmark for Dynamic Spatial Indexing. In *Mobile Data Management*, pages 92–105, 2003. DOI: 10.1007/3-540-36389-0_7. 15

[116] J. Mylopoulos. Conceptual modelling and telos. *Conceptual Modeling, Databases, and Case - An integrated view of information systems development*, page 49–68, 2008. 8

[117] F. Naini, O. Dousse, P. Thiran, and M. Vetterli. Population size estimation using a few individuals as agents. In *ISIT*, pages 2499–2503. IEEE, 2011. 111

[118] L.-V. Nguyen-Dinh, W. G. Aref, and M. F. Mokbel. Spatio-temporal access methods: Part 2 (2003 - 2010). *IEEE Data Eng. Bull.*, 33(2):46–55, 2010. 15

[119] T. Nicolai, N. Behrens, and E. Yoneki. Wireless rope: Experiment in social proximity sensing with bluetooth. In *PerCom*, 2006. 111

[120] G. Ozsoyoglu and R. T. Snodgrass. Temporal and Real-Time Databases: A Survey. *IEEE Transactions on Knowledge and Data Engineering*, 7(4):513–532, 1995. DOI: 10.1109/69.404027. 19

[121] J. Paek, J. Kim, and R. Govindan. Energy-efficient rate-adaptive gps-based positioning for smartphones. In *8th international conference on Mobile systems, applications, and services (MobiSys)*, pages 299–314. ACM, 2010. DOI: 10.1109/69.404027. 88

[122] A. Pantelopoulos and N. G. Bourbakis. A survey on wearable sensor-based systems for health monitoring and prognosis. *IEEE Transactions on Systems, Man, and Cybernetics, Part C: Applications and Reviews*, 40(1):1–12, 2010. DOI: 10.1109/TSMCC.2009.2032660. 55

[123] C. Parent, S. Spaccapietra, and H. Stuckenschmidt. *Ontology Modularization*. Springer Verlag, 2008. 17

[124] C. Parent, S. Spaccapietra, and E. Zimanyi. *Conceptual Modeling for Traditional And Spatio-Temporal Applications*. Springer Verlag, 2006. 8, 18

[125] B. W. Parkinson and J. J. Spilker. *Global Positioning System: Theory and Applications (volume One)*, volume 1. Aiaa, 1996. DOI: 10.2514/4.866395. 3

[126] N. Pelekis, Y. Theodoridis, S. Vosinakis, and T. Panayiotopoulos. HERMES - A Framework for Location-Based Data Management. In *EDBT*, pages 1130–1134, 2006. DOI: 10.1007/11687238_75. 14

[127] H. Peng, F. Long, and C. Ding. Feature Selection Based on Mutual Information: Criteria of Max-Dependency, Max-Relevance, and Min-Redundancy. *IEEE Transactions on Pattern Analysis and Machine Intelligence*, 27(8):1226 –1238, 2005. DOI: 10.1109/T-PAMI.2005.159. 98

[128] A. Pentland. Looking at people: Sensing for ubiquitous and wearable computing. *IEEE Transactions on Pattern Analysis and Machine Intelligence*, 22(1):107–119, 2000. DOI: 10.1109/34.824823. 1

[129] G. Perrucci, F. Fitzek, and J. Widmer. Survey on energy consumption entities on the smartphone platform. In *VTC Spring*, pages 1–6. IEEE, 2011. 85

[130] D. Pfoser. Indexing the Trajectories of Moving Objects. *IEEE Data Engineering Bulletin*, 25(2):3–9, 2002. 15

[131] M. Potamias, K. Patroumpas, and T. Sellis. Sampling Trajectory Streams with Spatiotemporal Criteria. In *SSDBM*, pages 275–284, 2006. DOI: 10.1109/SSDBM.2006.45. 29

[132] B. Priyantha, D. Lymberopoulos, and J. Liu. LittleRock: Enabling Energy Efficient Continuous Sensing on Mobile Phones, Feb. 2011. DOI: 10.1145/1791212.1791285. 61, 87

[133] C. Qin, X. Bao, R. Roy Choudhury, and S. Nelakuditi. Tagsense: a smartphone-based approach to automatic image tagging. In *MobiSys*, pages 1–14. ACM, 2011. DOI: 10.1145/1999995.1999997. 111

[134] M. A. Quddus, W. Y. Ochieng, and R. B. Noland. Current Map-Matching Algorithms for Transport Applications: State-Of-The Art and Future Research Directions. *Transportation Research Part C: Emerging Technologies*, 15(5):312–328, 2007. DOI: 10.1016/j.trc.2007.05.002. 28, 42

[135] L. R. Rabiner. A Tutorial on Hidden Markov Models and Selected Applications in Speech Recognition. *Readings in speech recognition*, pages 267–296, 1990. DOI: 10.1109/5.18626. 45

[136] K. K. Rachuri, C. Mascolo, M. Musolesi, and P. J. Rentfrow. Sociablesense: exploring the trade-offs of adaptive sampling and computation offloading for social sensing. In *MobiCom*, pages 73–84. ACM, 2011. DOI: 10.1145/2030613.2030623. 62, 89

[137] A. Rai, Z. Yan, D. Chakraborty, T. K. Wijaya, and K. Aberer. Mining complex activities in the wild via a single smartphone accelerometer. In *Proceedings of the Sixth International Workshop on Knowledge Discovery from Sensor Data*, SensorKDD, pages 43–51, 2012. DOI: 10.1145/2350182.2350187. 9, 54

[138] R. K. Rana, C. T. Chou, S. S. Kanhere, N. Bulusu, and W. Hu. Ear-Phone: an End-To-End Participatory Urban Noise Mapping System. In *IPSN*, pages 105–116, 2010. DOI: 10.1145/1791212.1791226. 111

[139] N. Ravi, N. Dandekar, P. Mysore, and M. L. Littman. Activity Recognition from Accelerometer Data. In *AAAI*, pages 1541–1546, 2005. 3, 53, 61, 69

[140] F. Reitsma. Modeling with the Semantic Web in the Geosciences. *IEEE Intelligent Systems*, 20(2):86–88, 2005. DOI: 10.1109/MIS.2005.32. 20

[141] D. Riboni, L. Pareschi, L. Radaelli, and C. Bettini. Is ontology-based activity recognition really effective? In *PerCom Workshops*, pages 427–431. IEEE, 2011. DOI: 10.1109/PERCOMW.2011.5766927. 8

[142] Å. Rudström, M. Svensson, R. Cöster, and K. Höök. Mobitip: Using bluetooth as a mediator of social context. *Ubicomp*, 2004. 3

[143] O. Saukh, D. Hasenfratz, A. Noori, T. Ulrich, and L. Thiele. Route selection for mobile sensors with checkpointing constraints. In *PerCom Workshop*, pages 266–271, 2012. DOI: 10.1109/PerComW.2012.6197492. 88

[144] D. P. Siewiorek, A. Smailagic, J. Furukawa, A. Krause, N. Moraveji, K. Reiger, J. Shaffer, and F. L. Wong. Sensay: A context-aware mobile phone. In *ISWC*, volume 3, page 248, 2003. DOI: 10.1109/ISWC.2003.1241422. 4

[145] L. Smith and M. Hall. Feature Subset Selection: a Correlation Based Filter Approach. In *ICONIP*, 1997. DOI: 10.1002/tee.20641. 70, 76

[146] S. Spaccapietra, C. Parent, M. L. Damiani, J. A. de Macedo, F. Porto, and C. Vangenot. A Conceptual View on Trajectories. *Data and Knowledge Engineering*, 65:126–146, 2008. DOI: 10.1016/j.datak.2007.10.008. 8, 16, 20, 25

[147] M. Srivastava, T. Abdelzaher, and B. Szymanski. Human-centric sensing. *Philosophical Transactions of the Royal Society A: Mathematical, Physical and Engineering Sciences*, 370(1958):176–197, 2012. DOI: 10.1098/rsta.2011.0244. 5

[148] S. Staab and R. Studer. *Handbook on ontologies*. Springer, 2009. DOI: 10.1007/978-3-540-92673-3. 18

[149] T. Starner, S. Mann, B. Rhodes, J. Levine, J. Healey, D. Kirsch, R. W. Picard, and A. Pentland. Augmented reality through wearable computing. *Presence: Teleoperators and Virtual Environments*, 6(4):386–398, 1997. 1

[150] D. D. Steinberg and L. A. Jakobovits. *Semantics: An interdisciplinary reader in philosophy, linguistics and psychology*. CUP Archive, 1971. 7

[151] M. Steyvers and T. Griffiths. Probabilistic topic models. *Handbook of latent semantic analysis*, 427(7):424–440, 2007. DOI: 10.1145/2133806.2133826. 77

[152] T. Strang and C. Linnhoff-Popien. A context modeling survey. In *Ubicomp Workshop Proceedings*, 2004. 8

[153] V. Subbaraju, A. Kumar, V. Nandakumar, S. Batra, S. Kanhere, P. De, V. Naik, D. Chakraborty, and A. Misra. Conferencesense: monitoring of public events using phone sensors. In *Ubicomp*, pages 1167–1174. ACM, 2013. DOI: 10.1145/2494091.2499775. 52

[154] T. Teixeira, G. Dublon, and A. Savvides. A survey of human-sensing: Methods for detecting presence, count, location, track, and identity. *ACM Computing Surveys*, 5, 2010. 5

[155] Y. Theodoridis. Ten benchmark database queries for location-based services. *Comput. J.*, 46(6):713–725, 2003. DOI: 10.1093/comjnl/46.6.713. 15

[156] R. P. Troiano, D. Berrigan, K. W. Dodd, L. C. Mâsse, T. Tilert, M. McDowell, et al. Physical activity in the united states measured by accelerometer. *Medicine and science in sports and exercise*, 40(1):181, 2008. DOI: 10.1249/mss.0b013e31815a51b3. 3

[157] M. A. Viredaz, L. S. Brakmo, and W. R. Hamburgen. Energy management on handheld devices. *Queue*, 1(7):44, 2003. DOI: 10.1145/957717.957768. 87

[158] X. Wang, X. Wang, and J. Zhao. Impact of Mobility and Heterogeneity on Coverage and Energy Consumption in Wireless Sensor Networks. In *ICDCS*, pages 477–487, 2011. DOI: 10.1109/ICDCS.2011.17. 88

[159] X. H. Wang, D. Q. Zhang, T. Gu, and H. K. Pung. Ontology based context modeling and reasoning using owl. In *PerCom*, pages 18–22, 2004. DOI: 10.1109/PERCOMW.2004.1276898. 8

[160] Y. Wang, Q. Jacobson, J. Lin, J. Hong, N. Sadeh, M. Annavaram, and B. Krishnamachari. A framework of energy efficient mobile sensing for automatic user state recognition. In *MobiSys*, pages 179–192, 2009. DOI: 10.1145/1555816.1555835. 62

[161] J. Weppner and P. Lukowicz. Collaborative crowd density estimation with mobile phones. In *SenSys*, 2011. 3

[162] K. Xie, K. Deng, and X. Zhou. From Trajectories to Activities: a Spatio-Temporal Join Approach. In *LBSN*, pages 25–32, 2009. DOI: 10.1145/1629890.1629897. 45

[163] Z. Yan, D. Chakraborty, A. Misra, H. Jeung, and K. Aberer. SAMMPLE: Detecting Semantic Indoor Activities in Practical Settings Using Locomotive Signatures. In *ISWC*, pages 37–40, 2012. DOI: 10.1109/Iswc.2012.22. 9, 63

[164] Z. Yan, D. Chakraborty, C. Parent, S. Spaccapietra, and K. Aberer. SeMiTri: a framework for semantic annotation of heterogeneous trajectories. In *EDBT*, pages 259–270, 2011. DOI: 10.1145/1951365.1951398. 8

[165] Z. Yan, J. Eberle, and K. Aberer. OptiMoS: Optimal Sensing for Mobile Sensors. In *MDM*, pages 105–114, 2012. DOI: 10.1109/MDM.2012.43. 88

[166] Z. Yan, J. Macedo, C. Parent, and S. Spaccapietra. Trajectory Ontologies and Queries. *Transactions in GIS*, 12(s1):75–91, 2008. DOI: 10.1111/j.1467-9671.2008.01137.x. 8, 16

[167] Z. Yan, C. Parent, S. Spaccapietra, and D. Chakraborty. A Hybrid Model and Computing Platform for Spatio-semantic Trajectories. In *ESWC (1)*, pages 60–75, 2010. DOI: 10.1007/978-3-642-13486-9_5. 8

[168] Z. Yan, V. Subbaraju, D. Chakraborty, A. Misra, and K. Aberer. Energy-Efficient Continuous Activity Recognition on Mobile Phones: An Activity-Adaptive Approach. In *ISWC*, pages 17–24, 2012. DOI: 10.1109/ISWC.2012.23. 62, 105, 106

[169] J. Yang. Toward physical activity diary: motion recognition using simple acceleration features with mobile phones. In *IMCE*, pages 1–10. ACM, 2009. DOI: 10.1145/1631040.1631042. 57

[170] J. Yang, H. Lu, Z. Liu, and P. Boda. Physical Activity Recognition with Mobile Phones: Challenges, Methods, and Applications. In *Multimedia Interaction and Intelligent User Interfaces*, pages 185–213, 2010. DOI: 10.1007/978-1-84996-507-1_8. 3, 52, 58, 61

[171] M. Youssef and A. Agrawala. The horus wlan location determination system. In *Proceedings of the 3rd international conference on Mobile systems, applications, and services*, pages 205–218. ACM, 2005. DOI: 10.1145/1067170.1067193. 13

[172] J. Zhang and M. F. Goodchild. *Uncertainty in Geographical Information*. CRC, 1 edition, 2002. DOI: 10.4324/9780203471326. 27

[173] Y. Zheng, W.-K. Wong, X. Guan, and S. Trost. Physical activity recognition from accelerometer data using a multi-scale ensemble method. In *IAAI*, 2013. 54

[174] Y. Zheng and X. Zhou, editors. *Computing with Spatial Trajectories*. Springer, 2011. DOI: 10.1007/978-1-4614-1629-6. 15

Authors' Biographies

ZHIXIAN YAN

Zhixian Yan works as a research staff member at Samsung Research America, located in San Jose, California, USA. Before joining Samsung, he was a postdoctoral researcher at the Swiss Federal Institute of Technology - Lausanne (EPFL) in Switzerland, where he also received his Ph.D. in Computer Sciences in 2011. His primary research interests include mobile sensing, mobile data management and data mining, wearable and mobile computing, semantic web, and web services. He has published over 30 peer-reviewed papers in top conferences and journals. He has received three best paper awards (including nominations or honorable mention) at MobiDE'12, ISWC'12, and MDM'12, respectively. He serves on the technical committee of several intentional conferences/workshops and journals. The work for this book was conducted while Zhixian was affiliated with EPFL.

DIPANJAN CHAKRABORTY

Dipanjan Chakraborty is a Senior Researcher at IBM Research India. He has over 14 years of experience in research and innovation. His traditional research interests are in mobile and context-aware computing and services, social networks, and sensor networks. Of late, he is focusing on mobile sensing and community-oriented sensing and analytics. To date, he has co-authored around 80 peer-reviewed publications and his work has received around 3000 citations. He has filed 21 patents and 8 issued patents. At IBM, he has received multiple research division accomplishment awards for his work in the area of context-aware computing. He has been involved in organizing several international conferences and workshops in his area such as Mobile Data Management, Percom, Comsnets, and the International Conference on Pervasive Services.

Printed in the United States
by Baker & Taylor Publisher Services